Praise for *Wasteland*

A 2023 Book of the Year in the *Guardian* and the *New Yorker*

'The book comes alive in its descriptions of people and places . . . Franklin-Wallis writes stylishly about ugly things . . . interesting and sobering . . . His book should prompt serious discussion in boardrooms and parliaments'
The Economist

'*Wasteland* is so captivating. It is an unflinching account of the best and worst of us, related through the things we choose to discard. Franklin-Wallis has travelled extensively to tell the story of waste and the result is fascinating'
Literary Review

'In *Wasteland*, he tackles all elements of the effects of waste, from cities in India and Ghana to the banks of the river Thames in London. His eye for detail, honed over years of non-fiction writing, turns the abstract into the immediate – and the alarming . . . *Wasteland* isn't a comfortable read, but it's an important one'
New Scientist

'Full of arresting figures . . . Franklin-Wallis, the features editor at *GQ*, grounds his narrative in firsthand reporting . . . Oh, and one more thing. We all need to buy less stuff'
Private Eye

T0187078

'In this brilliant book, Franklin-Wallis goes into it up to his neck – so we don't have to! A gripping read that will anger as much as it fascinates'
Hugh Fearnley-Whittingstall

'An urgent, probing and endlessly interesting investigation into our staggering wastefulness and the environmental crisis this is creating right under our noses . . . Tirelessly reported, it is a book both horrifying in its implications and gleefully hair-raising in the way it is told'
Cal Flyn, author of *Islands of Abandonment*

'This is an incredible journey into the world of rubbish, full of fascinating characters and mind-bending facts. My relationship with garbage is never going to be the same'
Oliver Bullough, author of *Moneyland*

'With his investigative chops and contagious curiosity, Oliver Franklin-Wallis has cracked wide a dozen hidden, jaw-dropping worlds . . . Yet despite its grim revelations, the book offers hope . . . *Wasteland* is compelling, smart, fair, often funny, always interesting, and just very important. Truly, it's the most impressive non-fiction I've read in quite some time'
Mary Roach, author of *Stiff* and *Gulp*

'A fascinating, deeply researched and hugely important exposé of what happens to the stuff we no longer want, and the social and environmental cost of dealing with it. Revelatory, thoughtful and honest about our complex relationship with waste'
Gaia Vince, author of *Nomad Century*

'Superb. Oliver Franklin-Wallis' deep dive into our wasteful ways and dirty histories turns up a story that gleams with insight and promise. An urgent and vividly told exploration of the underside of modern life, *Wasteland* also reveals what a better future could look like. You'll never see trash the same way again'
David Farrier, author of *Footprints*

'Wise, honest and unsparing, *Wasteland* will open your eyes to the reality of our throwaway society'
Henry Mance, author of *How to Love Animals*

'A book that wills you see the world quite differently than you did before'
Bill McKibben, author *The End of Nature*

'[A] scary reflection of our overconsumption and failure to deal with its impact . . . completely engrossing . . . it is the wake-up call we need to do better'
Euronews

WASTELAND

The Dirty Truth About What We Throw Away, Where It Goes, and Why It Matters

OLIVER
FRANKLIN-WALLIS

**SIMON &
SCHUSTER**

London · New York · Sydney · Toronto · New Delhi

First published in Great Britain by Simon & Schuster UK Ltd, 2023
This edition published in Great Britain by Simon & Schuster UK Ltd, 2024

1 3 5 7 9 10 8 6 4 2

Simon & Schuster UK Ltd
1st Floor
222 Gray's Inn Road
London WC1X 8HB

Simon & Schuster: Celebrating 100 Years of Publishing in 2024

www.simonandschuster.co.uk
www.simonandschuster.com.au
www.simonandschuster.co.in

Simon & Schuster Australia, Sydney
Simon & Schuster India, New Delhi

A CIP catalogue record for this book is available from the British Library

Paperback ISBN: 978-1-3985-0547-6
eBook ISBN: 978-1-3985-0546-9

Typeset in Perpetua by M Rules

Printed and Bound in the UK using 100% Renewable
Electricity at CPI Group (UK) Ltd

MIX
Paper | Supporting
responsible forestry
FSC
www.fsc.org   FSC® C171272

For Hannah

CONTENTS

THE TIPPING FLOOR

An alarm sounds. Blockage cleared, the line at Green Recycling in Maldon, Essex, rumbles back into life. A great churning river of rubbish rolls down the conveyor: Amazon boxes, splintered skirting board, crushed plastic bottles, sundry packets, newspapers by the hundred. I'm standing three storeys up on a green health and safety gangway, looking down the line. The air is thick, sour, the incessant roar of the machinery punctuated by the shattering of glass. It's an awful sight – and yet I'm mesmerised. Odd bits of junk catch my eye: a single discarded glove; a crushed Tupperware container, the meal inside uneaten; a lone photograph of a child, smiling atop an adult's shoulders. But they're gone in an instant. The line at Green Recycling handles 12 tonnes of waste an hour, and it's almost lunchtime.

Green Recycling is a Materials Recovery Facility, or MRF. When your recycling truck finishes its rounds on collection day, its destination will likely be a facility like this one, a huge corrugated-steel building on the end of an industrial estate, not far from the sea. Outside, trucks filled with refuse queue up to dump their loads in pre-marked bays. Below us, on the tipping floor, a man in an excavator grabs clawfuls of trash from towering heaps and drops it into the maw of an industrial bag shredder, which tears the bags open and spreads their contents across the conveyor. Along the belt, women in hard hats and acid yellow high-visibility vests deftly pick and channel valuables (bottles, cardboard, aluminium cans) into sorting chutes, which break off from the line like tributaries. The flow is relentless, the choreography balletic. It's a production line, in reverse.

'We produce 200 to 300 tonnes a day,' Jamie Smith, Green Recycling's general manager, yells over the din. Jamie's a career waste man, forty years old, with mud on his boots, dark hair and a strong, square jaw. He knows this line, its tics and its rhythms. Every day he watches as the vast output of humanity rolls by on its way to the end.

'Our main products are paper, cardboard, plastic bottles, mixed plastics, and wood,' Jamie explains, loudly, as we climb up a staircase to the sorting line. Like most in the business, he pronounces MRF as 'merf'. The conveyor passes through at hip height, the trash rolling by like a sushi buffet. None of the sorters raise their eyes. ('Females make the best pickers,' Jamie says. 'I don't know why, just do.') To my untrained senses the flow is indistinguishable, a nonstop torrent of brown and beige. 'We've had a big rise in boxes in the last couple of years, thanks to Amazon,' Jamie says, as if anticipating my thoughts. To the pickers, every item has an exact value, set that morning by a computer in the main office which tracks the international market for recyclables. Aluminium is highly prized, as are plastic bottles (preferably clear), 'though they don't much look like bottles by the time they get to us,' Jamie says. He picks a crushed one out of the flow with thumb and forefinger, dangles it appreciatively, and puts it down a collection chute. Each is only worth perhaps a hundredth of a penny – but with enough volume, one can make a decent living. And there's never a shortage of trash. 'The business is all about turning straw into gold,' Jamie says. A *Rumpelstiltskin* reference. I smile politely, struggling to see fairy tales in the dirt.

By the end of the line, the torrent has become a trickle, and the waste removed from the conveyor is stacked neatly in bales, ready to be loaded onto trucks. From here, it'll go – well, that's where it gets complicated.

It's the spring of 2019. I've come to Green Recycling to report a magazine story for *The Guardian* about a crisis in the waste industry, but as torn-up boxes and crushed furniture and packets in every conceivable form rolled past by the tonne, I realise that I've got the story wrong. Waste *is* the crisis.

Every day, our social media feeds light up with new evidence of the damage waste is wreaking on our planet: sea turtles caught up in six-pack rings, whales washing ashore with stomachs full of plastic. Each week it seems a new scientific paper lands in my inbox announcing that researchers have found waste somewhere new – microplastics in the soil, in the air we breathe, even in our bloodstream. Plastic waste is turning up in the melting glaciers of Everest[1] and in our deepest ocean trenches.[2] The Great Pacific Garbage Patch, the gyre that collects much of the estimated 11 million tons[3] of plastic dumped in the oceans every year, is now three times the size of France.[4] The problem is no longer even limited to the surface of the Earth. There's so much waste in orbit – detritus from old rocket launches, castoffs from the International Space Station, even one of Elon Musk's Teslas – that the European Space Agency is working on plans for an orbital clean up mission, in case the cloud of trash hurtling around the planet puts an explosive end to future space missions. This hypothetical event, known as Kessler syndrome, predicts that unless we act soon, human spaceflight will be grounded by what is, in effect, space littering.[5] At this point it will not surprise you to learn that when the last Apollo astronauts lifted off from the surface of the moon, they left their trash behind.

Human beings have always produced waste, but never before at such scale. Worldwide, we produced 2.01 billion tonnes of solid waste in 2016, the last year for which reliable data is available.[6] In the UK, the average person generates 1.1kg of waste every day; in the US, the world's most wasteful nation, the number is an astonishing 2kg per day.[7] The richer you are, the more you waste, and so as the developing world grows richer, the problem is accelerating. It is forecast that by 2050 we will be producing a further 1.3 billion tonnes a year, much of that in the Global South. Even so, 2 billion people currently live without access to solid waste collection, and fully a third of the world's solid waste is disposed of in what the World Bank calls 'an environmentally unsafe manner' – that is, it's dumped or burned in the open air. Inevitably, much of that waste is blown and washed into our rivers and seas, where it joins the toxic outfall from sewers, factories, power stations, and other pollution sources, so that many of the world's greatest rivers, from the Mekong to the Ganges, are increasingly hostile to living things.

None of this is news. We all encounter the waste crisis every day in our small ways, in our hedgerows and motorway verges, in our trees and gutters. But we've become so used to it that the scale of the problem can be difficult to comprehend. Take one everyday item of trash, a plastic bottle. More than 480 billion plastic bottles are sold worldwide every year – approximately 20,000 every second.[8] Stretched out end to end, they would circle the globe more than twenty-four times. And that's just one household item. (It's not even the most numerous. That dubious honour goes to the four trillion plastic cigarette filters flicked to the ground and stamped out annually.)[9] Now think about everything you touch in a typical day. We may occasionally consider – and in some cases take great care in knowing – where the things we use come from: where our food is grown, our clothes made, our iPhones assembled. But how many of us think about where it goes when we're done?

You drink a Coca-Cola, throw the bottle in the bin, put the bins out on collection day, and forget about it – but it doesn't disappear. From the moment the rubbish truck pulls away, your castoffs become the property of the waste industry, a vast global enterprise determined to extract the last penny of value from what remains. It starts with sorting facilities like Green Recycling. From there, your waste enters a labyrinthine network of brokers and traders. Paper is sent to mills, metal to foundries, glass washed and reused or melted down and remade. Food may be composted into fertiliser. What can be salvaged whole – clothes, phones, furniture – will be sold on, for another life in the thriving global second-hand market. If you're lucky (that is, if you live in a rich country) the chances are that your bottle will be loaded onto a container ship and shipped thousands of miles, to a recycling facility in South East Asia or Eastern Europe, where it might be recycled into a toilet seat or a pair of designer trainers. Or, if you're less lucky, that bottle will find its way onto an illegal waste dump in Malaysia or Turkey, where poor waste pickers, often children, pick through mountains of Western waste.*

* Waste has many synonyms – rubbish, refuse, junk, garbage, trash. In fact, these terms originally referred to different things: garbage, at least in American English, refers to food and other biodegradable (or 'putrescible') waste, whereas trash refers to dry waste. Rubbish, its preferred British equivalent, can refer to anything from ashes to recyclables. Refuse includes both garbage and rubbish, while junk tends to refer to disused objects, or scrap. In this book, like most people, I use the terms interchangeably – partly because in most of the world they end up in the same places anyway.

It's a harsh truth that for decades much of what we thought was being recycled actually wasn't, and isn't. For decades, Western nations have offshored our trash to poorer countries where labour is cheap and environmental standards weakest, a phenomenon known as 'toxic colonialism'. Thanks to the work of investigative reporters and NGOs, we now know that much of the recycling that we've exported in recent decades was habitually burned or dumped at sea; the waste clogging up Asia's rivers is, at least in part, our own. Much of the rest is being produced by companies in the Global North selling our hyper-consumptive lifestyles to new markets. In the UK, less than half of all household waste is recycled[10] (and even that number is an exaggeration, on which more later). Worldwide, the figure is just 20 per cent.[11] The vast majority of our waste is still disposed of in the same way that human beings have disposed of it for millennia: it's buried, or it's burned.

The environmental impact of all this is stunning. Today the solid waste industry contributes 5 per cent of all global greenhouse gas emissions – more than the entire shipping and aviation industries combined.* As it decomposes, rubbish produces methane, a potent greenhouse gas that traps many times more heat than carbon dioxide. Landfills ooze leachate, a waste industry term for the noxious black or yellow sludge that forms from the putrefying rubbish. Leachate is a noxious smoothie of every chemical and by-product you can imagine – acids, heavy metals, polychlorinated biphenyls (PCBs), dioxins and other poisons and carcinogens, which can leak down into the water table or into rivers, and into our water supply. As a result, modern landfills are sealed in

---

* This number is likely to be considerably higher once you include food waste – estimated to contribute between 8 and 10 per cent of total emissions alone.

with thick foundations and plastic barriers which prevent (or at least delay) leakage, but also prevent the waste from naturally breaking down, meaning that the contents often just lie there, entombed in toxic coffins. Such safety measures are expensive, and so in the developing world, landfills are rarely sealed; worldwide, open dumping accounts for more than a third of all waste disposal.

Waste is by nature a dirty word. Humans are biologically wired to avoid disgust, to look away from filth and decay. And so the waste industry is hidden, its trucks travelling in the early hours or late at night, its facilities operating at the edges of our awareness. Green Recycling lies at the very end of an industrial estate, surrounded by sound-deflecting metal boards. A machine called an Air Spectrum pumps out the smell of cotton bed sheets to mask the odour. You could live two streets away and never realise it was there. What little interaction we do have with the waste business is through names embossed onto bins – names like SUEZ, Biffa, and Waste Management Inc., all publicly listed corporations, yet ones you'll rarely encounter in the pages of a newspaper. And yet, worldwide the solid waste industry is worth billions, even before you add in sanitation or liquid waste – the effluents, both human and animal, that we pump into our seas and rivers.

The waste industry has profited from our disgust. There's an old English saying, 'where there's muck, there's brass'. Or another: 'one man's trash is another's treasure'. There can be riches in the rubbish, for those unashamed to get their hands dirty. Historically, waste disposal has been a business favoured by the unsavoury and the unscrupulous, from the New York mafia, to the Japanese Yakuza, to the Italian 'Ndrangheta, which in the 1980s and '90s illegally smuggled radioactive waste and dumped it off the coast of Africa.[12] Nobody really cares what happens to rubbish, so nobody asks questions – helpful, if what you're burying is bodies. But then rarely do we think about the people processing our trash, whether that's the garbage collectors doing their early morning rounds, the sorters at Green Recycling, or the more than 20 million people worldwide who make a living in the informal waste sector – painstakingly collecting and sorting through our discards.[13] These workers are more likely to be women, the elderly, or migrants, working in unhealthy and barely regulated conditions. We dump our waste on the margins, and on the marginalised.

Waste is history. Bodies decompose, paper moulds and crumbles, treasures can be pillaged or melted down by conquering forces. But nobody loots the rubbish tip, and so for centuries archaeologists, those ancient dumpster divers, have reconstructed our history from trash: discarded weapons, smashed pots and urns, food scraps with bite marks still notched in the bones. Waste can tell you a lot about a people: how they lived, how they ate, how they farmed and fought, loved and worshipped. It's by studying trash that we know the Maya first threw out their rubbish every month in what may have been early landfills, and that by around 500 BC, ancient Athenians had passed the first known sanitation law, ruling that household waste must be dumped at least a mile outside the city limits.[14] The Romans had trash collection of sorts, and were avid recyclers; recent excavation in Pompeii has uncovered sites where trash was sorted, and the salvageable material repurposed for the construction of new houses – a kind of Roman MRF.

Before the invention of the industrial economy, there was a lot less waste around. The biggest problem then wasn't trash but faeces, with its noxious odour and tendency to carry diseases. Even so, as towns grew into thriving cities, getting rid of all that waste soon became a problem. By the Middle Ages, the citizens of Paris were dumping so much waste outside the city walls that it was said an invading army could climb in over it – if it wasn't for the rats. The solution, invariably, was recycling. For centuries, informal garbage pickers were a central part of urban life. Almost nothing was wasted. Rags could be used as cleaning cloths, bones carved into cutlery or children's dolls, fireplace ash turned into bricks or fertiliser. Even the human excrement from privies was collected to be spread on the fields. By the nineteenth century, 'rag and bone men' were a common sight in Victorian London. So too were dustmen, mudlarks, toshers, pure-finders – all subcategories of scavengers, who collected everything from old brass to 'pure' (dog shit, for use in soap-making) and sold it on. In *Bleak House*, Charles Dickens wrote of the 'extraordinary creatures in rags secretly groping among the swept-out rubbish for pins and other refuse'.[15] It wasn't until 1875 that the British Parliament, responding to the growing social reform movement and a series of devastating cholera outbreaks, mandated that every household deposit their waste in a movable receptacle, which would be collected once a week – the invention of the modern dustbin.

Only after the Second World War did a revolutionary technology – plastic – begin to utterly reshape our relationship with waste. Plastic did more than generate huge new quantities of rubbish, it changed the way we talked and even thought about it. Until the invention of disposable nappies in 1943, the word *disposable* meant non-essential, a bonus – think of disposable income, which refers to the money left over after all our basic needs are already met. Suddenly, disposable had a new meaning: something designed to be thrown away. In 1955,

Life magazine reported glowingly about the rise of 'throwaway living'. 'Disposable items cut down on household chores,' it declared, below a picture of a joyous American family standing with open arms as these new disposable creations – plates, cutlery, pans, dog bowls, guest towels, you name it – rained down on them from above. (In a bitter irony, a recent study published in the journal *Science* revealed that tons of microplastic particles now literally rain down on us every year, a phenomenon the researchers termed 'plastic rain'.)[16]

The waste industry didn't just clean up after the modern economy – it enabled it. Corporations previously incentivised to produce high-quality products that lasted as long as possible were now lured into producing cheaper goods in ever greater volume, knowing that the consequences would fall not on their bottom line, but on the consumer. Along the way, the booming marketing industry gave us the concept of 'planned obsolescence', in which new products were designed to fail and thus need replacing – culminating in our modern world, where technology from smartphones to tractors can in many cases no longer be repaired without voiding the manufacturer's warranty. Today, one third of what we throw away is something produced the same year; and between 1960 and 2010, the amount of waste the average American was creating every year tripled.[17] The modern economy is built on trash.

By the 1980s Western countries were inundated with waste, so the industry did what the newly globalised economy was designed to: it offshored the problem. Every day, container ships arrived from China laden with goods for the Western market, and set off again largely empty. Soon, they were stuffed to the brim with trash. Between 1988 and 2018, nearly half – 47 per cent – of all plastic waste exported worldwide was sent to China for recycling.

Then, in 2018, China shut its doors. Under a sweeping new policy, 'National Sword', the Chinese government prohibited almost all foreign waste from entering the country, arguing that what was coming in was too contaminated, and that the environmental damage was too great. National Sword sent shock waves through the global waste industry. Prices for plastic, cardboard, clothing, and many other recyclables plummeted overnight. Recycling companies the world over collapsed.

It was that crisis that had brought me to Green Recycling. 'The price of cardboard has probably halved in twelve months,' Jamie told me. He looked dejected. 'It's hard. The price of plastics has plummeted to the extent that it isn't worth recycling. If China doesn't take plastic, we can't sell it.' Recently, he'd had to send several tonnes of recyclables that would have fetched good money on the open market to the incinerator.

After National Sword, exported trash began flooding into any country that would take it: Thailand, Indonesia, Vietnam. These countries all have something in common – they have among the highest rates of waste mismanagement in the world. Rubbish was dumped or burned in open landfills, or sent to recycling facilities with inadequate reporting, making the waste's final fate difficult to trace. Some Chinese businesses set up smuggling operations to get around National Sword and bring waste into China; as so often with waste, organised crime was involved. Local waste systems were quickly overwhelmed, and illegal dumping sites proliferated. Nobody wants to be the world's dumping ground, and so in the years since National Sword, countries such as Thailand, India, and Vietnam have all passed bans on the import of plastic waste. And yet still the rubbish flows.

* * *

When I first started my journey into the waste business, my initial feeling was not one of outrage, but of guilt. Like most people, I rarely thought about where my rubbish was going. I had naively assumed – what with our coterie of colour-coded bins and convoluted recycling logos on every packet – that waste was one of those issues that modern life had more or less solved. Then my wife Hannah and I had the first of our two daughters, and were stunned (and more than a little disgusted) at the sheer volume of waste this tiny new person could produce. We weren't zero-waste obsessives. I had a reusable water bottle and coffee cup, and a cupboard full of old totes to carry our shopping, but we used disposable nappies and wet wipes, ordered takeaways, and took our shampoo and toothpaste, like most people, from the inevitably plastic bottle. We were, I thought, rather average. But once I started thinking about waste, I couldn't stop. I started seeing it everywhere. I wanted to understand what had happened, how we got here, and what – if anything – could be done. *Wasteland* is my account of that journey.

This is not a book about rubbish. Though I started this journey with the story of trash and the people that dispose of it, I soon realised our real waste crisis is far bigger – a process that took me from landfills to ghost towns, from sewers to second-hand markets. Engineers and policymakers have historically considered solid waste and sanitation (that is, liquid waste) separate problems, the responsibilities of separate agencies and budgets. However, spend any time in the Global South, where streets flood with sewage because the open drains are clogged with trash, and you'll realise that the two are inextricably connected.

This is not a book about waste in both senses of the word, about what we throw away, but also the opportunities lost through our profligacy. Fully a third of the all the food we produce worldwide is wasted, and yet 820 million people go hungry every day. We've plumbed running water and sewage across most of the world, but thanks to climate change, we now face cities running dry. Tackling our waste crisis doesn't just mean removing litter from our rivers and oceans. Rethinking what we throw away could help us feed and water the world – and play a small part in saving our fragile planet.

Waste is not the most appealing subject for a book. When I first told people that I was writing about waste, they mostly pulled faces, or sounded at best bemused. Why would you want to write about something so *gross*? And yet, as I first saw on that morning in Essex, there's something beautiful, and even profound, about waste. There are few places in the world that give you a better view of humanity than a dump. I have seen bailiff's notices pass by on the conveyors, family photographs, and once, an opened letter from a child bereavement clinic. There are stories in all our discarded things: who made them, what they meant to a person before they were thrown away. The economist Leonard E. Read once wrote that no single person alive could make a pencil from start to finish; every step, from logging the timber, mining the graphite, machining the thing – hell, machining the machine – is built upon countless invisible hands. You could never visit every mine, every farm, every factory. But in the end, it all ends up in the same place – the endless ingenuity of humanity in one filthy, fascinating mass.

It's also easy to be disheartened by the waste crisis. Like climate change, it feels like a problem that individual action alone can do little to prevent. And yet, along the way, the more I met people determined to make a difference – farmers, sewage workers, garbage pickers, each passionate to save the world, or at least some small patch of it – the more optimistic I felt.

'We don't call it waste. We call it materials,' Jamie told me, that morning in Essex. To him, the endless river of trash wasn't a tragedy. It was an opportunity, each tossed item not an end, but a beginning of something else. As I set off on my journey, I hoped he was right.

PART ONE

DIRTY

I

THE MOUNTAIN

'You don't just fill the hole, you fill the sky as well'

The mountain looms over New Delhi, wreathed in smog. Even from a distance there's something uncanny about it, as if someone has lifted a chunk of the Himalayas and just dropped it here, on the eastern edge of India's capital city. It's January, cold, and a black cloud hangs over the mountainside as my rickshaw weaves through the crowded streets towards its base. Only as I get closer do I see that the cloud is in fact not a cloud, but birds: Siberian black kites and Egyptian vultures, thousands of them, careening over the mountainside like some dreadful murmuration. Scavengers. As the haze begins to clear, I see people silhouetted against the slopes, bent low beneath sacks twice their size. Dump trucks crawl up the long dirt road that winds to the summit. And then the smell hits: an acid tang that floods the nostrils and grabs at the throat.

This is Ghazipur landfill, a mountain not of stone, but of garbage — 14 million tonnes of it. Piled 65 metres high and covering an area of 28 hectares, it is the largest of three mega-landfills that ring Delhi.[1] The locals call it Mount Everest. 'Nearly taller than the Qutub Minar,' my guide, Anwar Ali, says, as we climb out of his dusty auto-rickshaw at the foot of the landfill. (The Qutub Minar, a thirteenth-century minaret, is one of Delhi's oldest and tallest landmarks.) A neat man in a red hoodie and suit trousers, with kind eyes and a sweep of dark hair, Anwar is a waste picker. He makes his living here on the mountain picking through the waste and salvaging what recyclables he can to sell. It's low work: dirty, dangerous, poorly paid, and yet all too common. Globally, as many as 20 million people make a living as waste pickers, predominantly in the Global South.[2] They go by many names: in Egypt, they are zabaleen, in Mexico, pepenadores, in Brazil, catadores, in South Africa, bagerezi. India, where they are sometimes called kabadiwala ('rag men'), is home to as many as 4 million waste pickers, around 5,000 of them working here on the mountain and living among the sprawling jhuggis, or informal settlements, around its base. However they are known, the waste pickers' task is fundamentally the same: they are the recyclers of the informal economy, finding value in what we throw away. In one man's trash, another's treasure.

I'd met Anwar and my other companion, an intense, thickly bearded man named Sheikh Akbar Ali, through the Delhi-based waste NGO Basti Suraksha Manch. Broad-chested in a puffer jacket and Covid mask, a woollen hat pulled tight on his head, Akbar is technically not a waste picker but a waste trader. When he's not making a living trading scrap plastics and metals, he volunteers his time helping the other pickers whenever they need help with healthcare or housing. The private company that owns the landfill is unwelcoming to journalists, so Anwar and Akbar have graciously offered to sneak me onto the mountain, to see it up close.

Ghazipur originally opened in 1985, as a place for New Delhi (then with a population of just 6 million) to dispose of its household waste. By 2002 the landfill had outgrown its original capacity, but with nowhere else to go, the waste kept coming – and so the mountain grew. Today, more than 30 million people[3] live in Delhi's greater metropolitan area, known as the National Capital Region. By 2028, it is expected to surpass Tokyo to become the world's most populous city. By the end of this decade, India's population, already an estimated 1.3 billion people, will overtake China's to become the world's largest. With so many people comes an unprecedented deluge of waste – at Ghazipur, another 2,500 tonnes arrive every day. And so as India's cities have grown, so have its landfills. (Ghazipur is not even the largest. That dishonour goes to the notorious Deonar landfill, in Mumbai.) Although in Ghazipur's case, *landfill* isn't really the appropriate term. It's better to call Ghazipur what it really is: a dump.*

If you want to tell the story of waste, there's no better place to start than a dump. For those of us lucky enough to live in a rich country, the very idea might seem anachronistic, something that went out of date along with the VHS. When I was a child, my parents would occasionally take my brothers and me to 'the dump', usually after Christmas or a garage clear-out. I remember the acid-sweet smell, the sound of shattering wine bottles, the illicit sense of permitted destruction. These days, in the UK, the closest most of us would get is the local recycling centre. But for much of the world, waste management still means one of two things: a landfill, or a dump. Worldwide, 37 per cent of our waste is landfilled, according to the World Bank; fully another third ends up on open dumpsites.[4] Waste management is expensive (all those garbage trucks, sanitation

---

* The International Solid Waste Association uses 'dump' to refer to waste disposal sites that lack the environmental controls (lining, topsoil cover) required of 'sanitary' landfills. That, of course, comes with problems – not least that it defines 'good' waste management as that which is sold by the Western waste industry – but for ease of use I'm sticking with it here.

workers and MRFs cost money) whereas dumping costs virtually nothing, so, as the global population has boomed, mega-dumps like Ghazipur have proliferated. Dumps with names like Bantar Gebang, Dandora, Olusosun, Matuail. And that is just in the cities. Travel through the countryside in almost any poor part of the world and you'll find dumps – on the edges of towns, in ditches, on the riverbanks and verges.

Dumps are as old as we are. Wherever human beings have walked, we have left waste behind: food scraps, fabric, parchment, broken tools. Our discards have allowed archaeologists to piece together our history as far back as the Palaeolithic. 'Waste is the secret history, the underhistory,' wrote Don DeLillo in *Underworld* – undoubtedly the finest novel ever written about garbage – but it is also just plain *history*. Middens, the name archaeologists give to bygone rubbish dumps, are one of our richest sources of insight into ancient civilisations all over the world. One of the largest sources of classical writing ever found is the rubbish dump at Oxyrhynchus, in Egypt, in which archaeologists discovered hundreds of thousands of documents, among them lost Christian texts and works by Plato, Euclid and Sophocles.[5] Ancient civilisations discarded so much waste that it accumulated in the soil, over time raising the level of entire settlements. These features, called tells, are found across the ancient world: whole cities, founded on rubbish.

'This is a treasure mine,' Akbar says. He means the dump. 'You can find anything, even 300- or 400-year-old things.' It happens all the time, he explains: people die and relatives clear their houses out, not knowing that what they've discarded is some priceless antique or piece of history. To them it is just trash. By the time it arrives here, caked in the filth of the dump, they're right.

In India, waste pickers are generally split into two categories:

those who collect waste on street carts to sort in the streets or at home, and those who work on the dumps. It is a gruelling task. At Ghazipur, the pickers set off before sunrise and pick until noon. The freshly tipped waste, which contains the best material, is dumped on the summit, which means a long climb. 'Winter, summer, even if it's raining, we will be drenched and still work,' Anwar says. In summer, temperatures in New Delhi can exceed 40°C. 'If it is 40 degrees right here, it would be 50 degrees up there, because of the heat from the waste as well as the sun,' he explains. To cope, the pickers hide bottles of water in the waste along their route, like climbers making camp. But for many of them the heat can be too much, and so they choose to work at night, by the light of head torches.

'When you see the waste, you see a dump,' Akbar says, gesturing towards the mountain. 'I do not see waste. I see the resources for myself. That waste is my livelihood.'

To reach Ghazipur's slopes, we must first cross a marshy pool of fetid wastewater. The pickers have made a bridge from stones and rubble bags. '*Rama Setu*,' Anwar says. He's referring to the Sanskrit epic *Ramayana,* which tells of how the Hindu god Rama built a bridge of stones from India to Sri Lanka in order to rescue his bride from a demon king. (The tale serves as an origin story for a real geological formation that once connected the two countries.) Anwar is joking – making light of their forced ingenuity – but it's also an acknowledgement that we are crossing a border, to somewhere hostile, other.

I scrabble up a dirt bank and onto the dump. On the western boundary, freshly laden trucks are lining up to begin the climb to the summit. The bird-cloud wheels angrily above them, waiting for butchers' scraps which arrive daily from nearby markets; the way they swerve and screech makes me wonder if they're repulsed, too. Not wanting to be caught, we skirt another route south up onto a low ridge. There's a wide road around the base, which slopes upwards. At its foot, leachate the colour of tar has carved rivulets in the waste, which run downhill into an ink-black catchment pond strewn with plastic. The stench is formidable, caustic and invasive. I can feel it in my sinuses like an oncoming migraine.

Every dump or landfill, I have come to learn, has a unique smell – the exact aroma differing based on variables including the exact constituents of the waste, the temperature, and the weather. Methane and carbon dioxide, the primary by-products of decomposing rubbish, are odourless. The rotting smells you most likely associate with garbage are hydrogen sulphide (the smell of rotting eggs) and dimethyl sulphide (rotting vegetables). But waste is a complex bouquet. Its top notes can include ammonia (urine), acetaldehyde (bad cider), and trimethylamine (rotting fish). Perhaps the most interesting ingredients are the diamines putrescine, the compound that gives rotting meat its odour, and cadaverine, so named for being the unique scent of rotting corpses. For once I'm grateful that, thanks to the pandemic, I have a facemask to hand, and pull it tightly over my nose.

Where the road through the dump has been dug, the trash has formed a sheer cliff face. Over time, the weight of the upper slopes has crushed the earlier deposits into a kind of novel sedimentary rock: clothing scraps, drink cans, packets of masala-flavoured crisps, Mom's Magic biscuits, ceramic shards, a chair, men's razors, women's razors, children's toys, all held together with a nameless brown grime, like one great capitalism sundae. Spend enough time on a dump, I have found, and waste starts to blend together, the way clouds do, until it is not many things but one thing – a foul, abstract mass. Looking at it, I'm reminded of class geography trips as a child, visiting cliffs and marking off epochs in the geological strata. This one might be marked *Anthropocene*.

We climb. The waste on the lower levels, compressed by weight and time, feels firm, but on the slopes it is loose and unpredictable, scree-like. Trash crunches loudly underfoot. 'It takes time to fill in, so that it becomes solid,' Anwar explains. During the wet season, rainwater trickles down between the waste, forming subterranean flows that can carve off larger sections, triggering landslides. Just a few months earlier, a chunk of the mountain gave way, killing two people and carrying a car into a nearby canal.[6] Such incidents are not uncommon: in 2017 alone, landfill collapses killed at least 150 people worldwide, including 29 when a dump in Colombo, the capital of Sri Lanka, collapsed onto a residential neighbourhood.[7]

'Have you ever been injured?' I ask Anwar.

'Yes,' he says, though never badly. He's seen others maimed, even killed. Over time, he explains, you learn to read the waste the way sailors can read a river's current; he can intuit what is likely to be solid, what isn't. But collapses are unpredictable. 'Nobody knows what is going to happen. A person is working until the end, when they go down.'

I notice small plumes of what looks like white dust here and there, caught in the wind. 'Those are fires,' Anwar says, casually. Flame is a constant danger on landfills, waste being to a large degree fuel. Fresh rubbish might contain hot ashes, unburnt wood, or batteries, which explode when heated or punctured. Decomposing waste creates heat, which can react with the landfill gases and spontaneously combust. Most fires are small and can be put out by the pickers. But others smoulder for days or weeks under the surface. Occasionally fires get out of control. A few weeks after I visit, a massive blaze will break out on the dump, burning for forty-eight hours before eventually being extinguished.

Such disasters have added to Ghazipur's cursed reputation. 'This is a black spot for Delhi,' Anwar says. 'On every border – with Haryana, Punjab, and Uttar Pradesh – there are landfills. So whenever a person comes to Delhi, the first thing they see is this.' In recent years, the municipal government has declared its intention to reduce the height of Ghazipur, and to eventually close the city's open dumps. It is not alone. Over the last decade India has undergone an extraordinary sanitation drive, led by its controversial prime minister, Narendra Modi. Upon his election in 2014, Modi launched a $30 billion campaign to improve sanitation in India's cities and villages. It was this, the Swachh Bharat, or 'Clean India' campaign, that had brought me to India – arguably the largest drive to clean a nation and eradicate waste since the Industrial Revolution. Modi's waste drive has coincided with an international effort, led by the World Bank and the International Solid Waste Association, to close open dumpsites and replace them with sanitary landfills, energy-from-waste plants, and other, less polluting disposal methods. In practice, this has meant the widespread privatisation of waste collection, with contracts usually going to large private waste companies, while placing the livelihoods of waste pickers under threat. 'They have completely pushed us out,' Akbar says.

Someone shouts, and we look up. On a ridge high above us a picker is hauling a massive load of waste wrapped inside a net. He yells a warning, and hurls the load off the cliff. It careens down the mountainside like a loose boulder, crashing to the floor below. Anwar motions to the ridge in question. 'This is new,' he says. Not long before my arrival, the city had levelled off Ghazipur's previous peak into a flat-topped mesa – less

Everest, more Table Mountain — and was in the process of reprocessing some of the waste into construction materials and refuse-derived fuel. But the reality is that the waste is simply spreading outwards, rather than up. The episode is the latest in a long history of negligence at Ghazipur. In 2015, the city opened an energy-from-waste plant on the edge of the landfill. However, the plant has been beset by problems. 'The factory has not been functioning properly these last two years,' Anwar says. Initially, many of the waste pickers signed up for jobs at the new plant. However, after what they allege was mismanagement and technical problems, most quit the facility. 'But the ownership wants to show that we are continuously working on a project that has taken hundreds of lakhs of rupees [that is, several million pounds] so a couple of times a week workers move waste inside and out,' Akbar says. We head to a vantage point overlooking the plant. The storage warehouse doors are open, with waste strewn everywhere. The incinerator sits silently, the stack spewing no vapours. There's nobody inside.

'Look,' Anwar says, pointing upwards. 'Kids.'

In the distance, three slight figures are coming back down with their hauls. Three girls, barely teenagers, are walking in concert, bags of trash balanced on their heads. I count at least a dozen children on the mountain, and more nearby, helping parents while they sort, or playing among the garbage. The state provides free schooling, but when they shut during the Covid pandemic, few of the children had laptops or smartphones to study from home, so most chose to go to work with their parents. 'They realise they can earn money at such a young age,' Anwar says, 'so that is why you see a lot of kids dropping out of school.'

The pickers are worried that we'll be spotted if we climb higher, so we head back to the clearing. Near the leachate pool,

I spot plants growing up through the garbage: cherry tomatoes, their green fruits sprouting in the winter sun. I try not to imagine what kind of chemicals their roots must be drawing up from this toxic soil. 'When these tomatoes grow to a significant size,' Anwar says, 'these waste pickers pluck them for dinner.'

Back in the clearing, a herd of cattle is roaming through the waste piles, grass-starved, picking at the dirt. Each pile belongs to a different family. Some are wrapped in woven polypropylene sacks, others simply bundled in blankets. I meet a man and his wife sat on a bright blue tarp, sorting through their haul: shampoo bottles, cans, cutlery, a children's plastic cricket bat, various toys. They sort them into bags, based on material. The man, Qasim,* has a thick moustache, wears a T-shirt and jeans and a bright orange cap, but no gloves, nor anything in the way of protection. Neither does his wife. Taking out a battered children's toy, Qasim cracks it open with a hammer, scooping out the insulated wiring as if shelling shrimp. Metals earn the highest prices; copper wire is particularly prized. Then it's higher-grade plastics – PET, HDPE – before you move on down. 'We have eighty-seven categories of waste,' Akbar tells me. Plastics, in particular, can be hard to tell apart with the naked eye – PET, HDPE, PVC, PP.[8] The pickers give them nicknames in their local languages: 'natural', 'BP', 'milk'. 'If I started teaching you right now, it would take you a year or more to get it,' Akbar says. Watching the pickers work, I don't doubt him.

A waste picker on Ghazipur might earn ₹7,000 per month, or around £2.30 per day, depending on the price of recyclables. Elsewhere in the city, where the kabadiwalas collect waste from the streets and households, earnings are slightly higher, although not by much. When the local plastic recycling

* Not his real name.

factories shut during the pandemic, prices crashed. Many of the pickers were forced to take out informal loans from the waste dealers, who charge high interest rates, trapping the pickers in a cycle of what is essentially debt bondage. The pickers couldn't even work, due to citywide curfews. 'The police would beat us up, and even if there was video evidence of it, no one would do anything,' Akbar says. Many waste pickers that I met in Delhi had similar stories.

It's not just in India. Globally, waste workers are often among the poorest and most marginalised in society. Few countries even recognise waste picking as a formal profession. 'After the waste is gone, we are like shit to them,' Akbar says. 'They don't like to look at us.' In India, Hindu waste workers are often Dalit – that is, deemed to belong to the lowest stratum of India's notoriously unjust caste system. Even among Dalits, landfill picking is low-status work; like 'manual scavenging' (cleaning out cesspits and sewers by hand) it was historically left to the most disadvantaged, the so-called 'untouchables'.

Here in Ghazipur, most of the waste workers are not Hindu but Muslims – migrants from poor and climate-stressed farming regions such as West Bengal and Uttar Pradesh. Others are economic migrants from the south, or from neighbouring Bangladesh. (All over the world, waste workers are disproportionately likely to be immigrants, trash picking being a job with among the lowest entry requirements.) In India, Muslims have faced growing persecution in recent years, amid an alarming rise in Hindutva nationalism. Recently, some Hindu politicians have openly called for violence, and the expulsion of Muslims from the country, something unthinkable even a decade ago. Akbar has seen the tide turning even among waste pickers. 'The difference between the politics of then and now,' Akbar says, 'is of the ground and the sky.'

A small crowd has formed while we've been talking, every-one intrigued as to what this strange, tall white man wants here on the mountain. A group of young girls are giggling beside someone I take to be their mother. They're dressed in long scarves and Western clothes; one wears a sweater, clearly counterfeit, printed with the Apple logo. Another notices a smashed digital camera, and they take turns pretending to take photos of each other, an imagined fashion shoot among the broken things.

'What do you want to be when you grow up?' I ask.

'Doctors,' one says. The others collapse into laughter.

Later, as I turn to go, their mother says, 'Take them with you.'

I would like to, I say, even though I know I can't. I think of my own daughters at home, and the all-engulfing despera-tion it must take to offer your child up to a perfect stranger. Akbar signals that it's time to leave. As we walk through the waste piles back to Anwar's rickshaw, two of the youngest kids find a half-crushed bottle of bubble mixture in the dirt. One unscrews the cap and is delighted to find it half-full. We drive away, their tiny bubbles bursting in the breeze.

Another day, another dump. This one 6,700km away, on England's wind-bitten north-east coast. I've not long passed Newcastle when I see the first sign of my destination: a queue of green articulated lorries, disappearing behind a thick treeline. I turn in to follow. Rain has fallen, and a concrete-grey cloud front is still unfurling overhead, the texture of mineral wool. Mud viscous, puddles deep. Wind turbines turn slowly on the horizon, surrounded by green fields and hedgerows. Beyond the fence-line I pass a weighbridge where the lorries are appraising their hauls and approach a cluster of drab temporary buildings. Ahead of me is not a mountain, but a series of low dirt hills, scattered with dump trucks and earthmovers. There isn't a person in sight.

If Ghazipur is among the worst that dumps have to offer, then this is the cutting edge: Ellington sanitary landfill, owned and operated by the multinational waste corporation SUEZ. It has taken me a long time to get here. In Britain, as in Europe, landfills have a lousy reputation as being archaic, foul and polluting. As a result, waste companies, always keen to show off their latest recycling facilities, are reluctant to let anyone on a landfill, and particularly not a journalist. In fact, for more than a year, every waste company that I ask refuses to grant me access to one, citing either the coronavirus pandemic or unspecified safety concerns (while at the same time, insisting that landfills are perfectly safe). SUEZ, the French waste and sanitation giant named for its involvement in digging the Suez Canal, is the one exception. It is more than happy to show me around.

In the office, I meet my chaperones: Victoria Pritchard, SUEZ's regional manager for the North East of England, and Jamie McTighe, the regional operations manager. Victoria – 'Vic' – wears a peroxide-blonde bob and a leopard print blouse under her high-vis jacket. Jamie is less glamorous but no less experienced, a nineteen-year waste man, taciturn in a hoodie and jeans. We're joined by Carolyn Fitzgerald, SUEZ's corporate affairs person, who is here to enthusiastically promote the company line.

Vic is warm and chatty, and not your typical landfill worker. 'I wanted to be a stage manager,' she says, laughing. She started at SUEZ as a personal assistant, and over the years has worked her way up through the ranks. 'When I got the supervisor job, they said to me, "But you're a *woman*? How are you gonna pick a bag of bentonite up?" Which is what, 25 kilos? "And how are you going to pull a pump out of a well?"' Vic says. 'I said, "Well, I'm not, because the clue is in the title – I'm going to supervise somebody else to do it."' Vic is now responsible for every SUEZ landfill stretching from here up to the Scottish borders. I like her immediately.

Landfill has always been a male-dominated industry. 'The ill-fitting PPE [personal protective equipment] I've had over the years is nobody's business,' Vic says, grimacing. 'We only get them in men's sizes. I don't want a pink one, it doesn't have to have flowers on. I just want something the same as everybody else that fits and does its job.' Vic is fastidious about landfills, insistent on them being correctly run and correctly named. She hates it when people call Ellington a 'tip'. 'I don't use the word tip. Because in the North East of England, a tip is a mess. My mum would say to me, "Your room's a tip!" It's not a tip. To me it's about the perception. My goal is you shouldn't know you're driving past a landfill. We should be operating in such a manner that you don't know you're on a landfill until you're tipping in your waste.'

Historically, landfills were often just holes in the ground – mines, bogs, marshes, deserts – any land on the fringes that people didn't have a use for. Waste was literally 'fill'. The practice of dumping rubbish and shaping it into new land goes back centuries: New York, Boston, and San Francisco, as well as parts of London, for example, are all built in part on trash. Robert Moses, the notorious city planner who reshaped New York during the middle decades of the twentieth century, was particularly fond of the landfill method: JFK and LaGuardia airports are built atop reclaimed waste, as is most of the south Manhattan shoreline, and Flushing Meadows, in Queens. In Britain, what we would now recognise as landfill didn't really take off until 1956, with the passage of the Clean Air Act. Prior to that, much of our waste was burned, either at home, or in municipal incinerators. But in 1952, the Great Smog – an air inversion which trapped the capital's noxious air pollution close to the ground for five days, killing an estimated 4,000 people – triggered a ban on the burning of household waste. The result was a vast increase in municipal garbage, which had to be sent to landfills.

Ellington is a sanitary landfill, meaning it is separated into standalone sections, or 'cells', which each contain around 250,000 tonnes. The first 'official' sanitary landfill opened in the city of Fresno, California in 1937,* the brainchild of Jean Vincenz, the city's then-commissioner of public works – though Vincenz, a keen sanitary reformer, had taken inspiration from similar methods employed earlier in England, where the practice was called 'controlled tipping'. Either way, sanitary landfills marked a minor revolution in waste management, introducing new methods to control both pests and odour. The Allied forces employed similar techniques during the Second World War, and after the war, returning soldiers quickly spread the practice throughout the land.

Today sanitary landfills are complex, highly engineered and regulated constructs. Even before ground can be broken, there must be geological surveys and environmental assessments – not too close to groundwater, or rivers, or anywhere that might disturb protected species. Only once approvals are met can the engineers begin to dig. Even the word landfill, it turns out, is a misnomer. 'You have landfill, landform or landraise,' Vic says.

---

* The Fresno landfill, which closed in 1987, is now a park, and a US site of historical interest. I've been there. It's a pretty place, a grassy hill overlooking acres of fruit fields and frequented by dog walkers.

Exactly which depends on location, and what the owners intend to use the land for later, after remediation. On the wall of one of the offices is an artist's impression of what Ellington will look like after it has closed. Vic and Jamie's job is to recreate it in dirt, like sculptors meeting a client brief. 'There's actually what we call "cell" and "air",' Vic says. 'You don't just fill the hole, you fill the sky to a certain point as well.'

Once it's on the landfill, the waste undergoes a biological and chemical onslaught. First, there are the pests: maggots, worms, insects, rats. Higher on the food chain, feral pigs, baboons and bears have all been found dining out on landfills, which to them must seem like a free buffet. Then there are the birds. (There are always birds.) In India, it is kites and vultures; in North America, it's starlings. In Spain, some white storks have become so reliant on the free, year-round supply of food from landfills that they have abandoned their seasonal migration. In Australia, the sacred ibis is so often seen rooting through trash that it earned the nickname 'bin chicken'.

Here in Britain, the main pests are gulls. Common gulls, herring gull, black-backed, black-headed – the waste attracts allcomers. This morning at Ellington they are amassed in a field just over the fence-line, an ominous feathery mob. SUEZ employs a full-time falconer, who uses birds of prey to keep them out. 'The main thing with them is varying techniques,' Jamie says. He points out a yellow box off in the distance, an electronic deterrent system. 'It'll start screaming like a bird, and that scares them off,' he says. At the prow of the hill, a falcon-shaped kite dangles from what looks like a fishing line, a kind of scare-gull.

Even before it arrives, the waste is already being devoured. When we talk about waste biodegrading, we're really talking about microbes: first aerobic bacteria, which break down the organic compounds (fats, proteins, carbohydrates) within the waste, setting loose sugars and amino acids, as well as carbon dioxide. Move deeper and anaerobic microbes take over, fermenting this chemical soup to produce a pungent mix of acids and alcohols. By now the landfill is quite literally a hot mess, producing both plumes of carbon dioxide and the afore-mentioned pungent gases – hydrogen sulphide, ammonia – as well as volatile organic compounds (VOCs) such as benzene. Finally, methanogenic bacteria take hold, and convert the acetic acid into methane. This final stage, which can happily continue within the soil for decades, is what makes the emissions from dumps so problematic for the climate. As well as being extremely flammable, methane is at least twenty-eight times more potent as a greenhouse gas than carbon dioxide, because it traps atmospheric heat more effectively.[9] All told, according to the US's Environment Protection Agency, land-fills are the third largest cause of methane emissions in the United States. Even that may be an underestimate. In recent years, aerial observation and multispectral satellite imaging has discovered that landfills produce far more greenhouse gases than previously thought. A recent aerial study by NASA's Jet Propulsion Laboratory found that several landfills in California were leaking methane at rates up to six times higher than EPA estimates.[10] The situation is worse in poorer countries, where methane and leachate from landfills and dumps is not collected. In 2021, the European Space Agency's Sentinel 5-P satellite detected a single landfill in Lahore, Pakistan, that was emitting 126 tons of methane an hour, putting its hourly emissions on par with that of about 6,200 cars.[11]

No landfill is the same. The actual constituents of the waste are as varied as what we throw away, which is to say that they can and do contain almost anything. Since the 1970s, most countries have separated out hazardous waste (hazardous waste landfills, in general, have stricter environmental controls, but much worse contents). But even your everyday municipal landfill is a haven of toxic and hazardous chemicals found in our cleaning products, makeup, pharmaceuticals, batteries and electronics. A recent study found that US landfills typically contain hundreds of toxic chemicals, ranging from agricultural pesticides to industrial pollutants, PCBs, Bisphenol-A, per- and polyfluoroalkyl substances (PFASs) – many of which are potent human toxicants, causing cancer, endocrinal disruption (that is, they mess with your hormones) and neurological damage.[12] Pharmaceuticals, including antibiotics, find their way in; so too do heavy metals like lead, mercury and zinc, which arrive inside toys, household electronics, lightbulbs, televisions and old-fashioned lead paint. Of the 1,333 sites currently on the US's Superfund National Priority List – the index of the most toxic places in America – 159 are former landfills.[13] No wonder, then, that working on or living near landfill sites is associated with a greater risk of asthma, respiratory disease, leukaemia, low birth rates and several types of cancer.[14]

To keep that noxious soup inside, each cell of a modern sanitary landfill is carefully engineered. 'I like to use the analogy of a pie,' Vic says. 'You have your bowl, and you rely on your bowl to make sure that all the inner of the pie doesn't leak out.' The hole is first lined with bentonite clay, which is then covered with a non-permeable plastic liner, welded shut at the seams. 'That's your crust.' The liner is topped with gravel and sheets of a hessian fibre, for drainage. Leachate within the landfill slowly trickles downwards, picking up chemicals. (It's this fluid that I had seen forming in pools at the base of the mountain at Ghazipur.) The actual chemical makeup of leachate depends on the waste. It has its own terroir, like wine. 'Like Burnhills [in Northumberland], where Jamie got his roots, was a special waste site, so their leachate is bright green,' Vic says.

'You'd get stuff in drums, and a lot of off-spec prescription medication,' Jamie explains.

'Go to Scotland, and at one of our sites the leachate produces this polymer,' Vic says, excited, taking out her phone to show me a picture. It's horrid: a black, spongy mass, like aerated chewing gum.

'It's a bit like whale blubber,' Jamie says.

'When you touch it, it goes *glublubglug*—' Vic wobbles her hand like set jelly, gleeful as a toddler.

At Ellington, the leachate is drawn out of the waste using a lattice of pipework and pumped to an onsite chemical treatment plant, before being sent to a nearby sewage plant for further processing. Vertical pipes draw off the landfill gas, which is used for power generation. The waste buried at Ellington currently produces around 1.3MW, or enough to power 3,500 homes. When a cell is filled, it will then be covered with another plastic liner and welded shut at the seams. Finally, the entire cell is covered in clay and several feet of soil, sealing the waste inside. The result is that modern landfills are less like dumps than balloons full of garbage.

The American archaeologist and anthropologist William Rathje made a career of studying landfills. In 1973, Rathje and his colleagues at the University of Arizona established the Garbage Project, which set out to examine waste with the same rigour that other scientists would apply to ancient archaeological dig sites. 'If we can come to understand our discards,' Rathje wrote, 'then we will better understand the world in which we live.'[15] They dubbed themselves garbologists. Trash, Rathje found, could tell you more about a neighbourhood – what people eat, what their favourite brands are – than cutting-edge consumer research, and predict the population more accurately than a census. Garbology has, for instance, shown that people consistently underestimate how much unhealthy food they consume, while overestimating their consumption of fruits and vegetables. Unlike people, garbage doesn't lie.

Rathje's researchers consistently found that landfills are less biologically active than assumed. Deep below the surface, the biological processes that break down the waste can slow to a standstill. After taking cores from 20-year-old landfills, Rathje and his researchers found that paper was still clearly legible, and even food waste was still easily identifiable: 'onion parings were onion parings, carrot tops were carrot tops. Grass clippings that might have been thrown away the day before yesterday spilled from bulky black lawn and leaf bags, still tied with twisted wire.'

And, inevitably, landfills leak. Liners are torn and ripped open by the waste, root systems and pests; plastic liners are broken by weathering or the chemicals within. Landfills built along coastlines or along waterways – a common practice for centuries – are constantly under threat from flooding and sea-level rise. Over time they erode, until inevitably they spill their contents. Trash eats its way out; nature finds a way in. Thus, as the writer Elizabeth Royte puts it, 'Over time, landfills pose more of a threat to the environment, not less.'[16]

We stand on a ridge and look out over the site. Where we're standing, Jamie says, is actually historic landfill – waste built upon waste. The empty cell falls away in front of us. Beyond it, previously filled cells rise in a series of gently undulating hills. On a ridge line atop the nearest cell, six articulated lorries are queuing up to dump their loads. But it's difficult to see, being so far away, so Jamie asks if I want to see the face up close. Of course I do, so we climb into the cab of a John Deere tractor and set off over the landfill. In the cab, we get to talking. Jamie grew up in Hull, he says, and studied environmental engineering at university. He had wanted to join the Environment Agency. This might seem the diametric opposite of that, but in fact, he says, it's the same thing – stewardship of the land. Managing a landfill is about preventing pollution, ensuring that the waste ends up in the ground. As we approach the open face, we pass two massive grates the size of billboards, designed to catch plastic bags and the like from blowing away in the wind. In heavy winds, the landfill closes. The fences that surround the site are less about keeping people out than waste in.

The ground underneath us is spongy, the waste still settling. Looking down, I notice that what looks like soil in fact isn't. We're driving over daily cover, a preliminary layer of material poured over the waste to fend off birds and lessen the odour. The dirt is flecked with brightly coloured plastics the size of confetti. 'Rejects from a plastic recycling process,' Jamie says.

At the active face, one of the lorries is still dumping its cargo – an anonymous clod of mattresses, drywall, and insulation. Construction waste. Not that it makes much difference to Jamie and the crews. 'Once it gets here, we can't do anything. It's all treated in the same way,' he says. The compactor, a massive 45-tonne machine with spiked metal wheels each the size of a small car, rolls over the newly deposited trash, crushing it down and kicking clouds of sawdust into the air. For every load, the compactor will make seven to ten passes, enough to squeeze out any gaps in the waste and make the mass more stable. Then another of Jamie's crew comes in, driving a track-loading shovel, and begins shaping the waste into the pre-agreed landform, in this case a gently sloping hill. Watching the operation for a while – dump, crush, spread, repeat – is mesmerising. The crew are not dumping the waste so much as sculpting it, carving a new landscape.

Jamie has worked on landfills long enough that he has seen people dump almost anything. I ask if that includes corpses (landfills being, statistically speaking, a common place to find murder victims), but he himself hasn't seen any. He did, however, once receive the 30ft carcass of a whale that had beached nearby. 'In those situations, it's the people who have to get it into the vehicle that I feel for,' he says, wincing. Jamie's crew dug the whale a grave and buried it, to protect it from the gulls – a small act of dignity within an otherwise foul end. Occasionally, landfills are host to even more exotic discards. In 2013, James Howells, a computer engineer in Newport, South Wales, accidentally threw out a memory card containing 8,000 Bitcoin, which ended up in the local landfill. At Bitcoin's peak, in 2021, the card's contents were worth nearly half a billion dollars. For the last few years, Howells has been trying to convince the city to dig up the landfill to find the card, offering to share the proceeds if found. So far, they have declined his offers – and so the riches remain buried.

Working on landfills, Jamie admits, 'isn't very glamorous'. The pay is mediocre, the conditions rarely pleasant. The operators spend almost all day in their cabs except for breaks, hauling and dumping whatever the weather – rain, shine, or snow. All of that moving bulk and heavy machinery can also be dangerous. In the US, garbage workers are almost three times as likely to be killed on the job than police officers.[17] Statistically, waste work is one of the most dangerous jobs there is.

'It is brutal out there, absolutely brutal,' Vic says.

The work is also increasingly hard to find. In the UK and

Europe at least, landfills are a dying business. (The same is not true in the US or Australia, which send 50 per cent and 30 per cent of waste to landfill, respectively.)[18] In 1996, the UK government introduced a landfill tax, to encourage recycling rates. It has steadily increased ever since; at the time of writing, landfill tax is at £98 a tonne. Anything that cannot be sent for recycling or composted is increasingly burnt inside energy-from-waste (EfW) plants. These days, the landfill only receives household waste from the local council when the nearby EfW is overflowing, or down for maintenance. The result has been a mass exodus from the landfill industry. 'It was a bit like rats leaving a sinking ship,' Jamie says. 'A lot of people left. The people that we have got left now have all been in the industry for quite a long time.'

The reason for landfill's decline is not, as was once feared, that we are running out of room for them (quite the contrary – empty land is easy to find). Rather they have become obsolete, unable to compete with recycling and energy generation on cost, as well as politically unpopular. Of the ninety-eight landfill sites that SUEZ operates in the UK, only four are still active – that is, accepting new waste. Victoria manages all of them. 'The landfill tax killed this industry too quick,' Vic says. Landfill operators, who run on thin margins, suddenly found themselves unable to fill the holes before their licences expired. One of SUEZ's sites was forced to write off £1 million. At the current rate, Vic doubts even Ellington will reach its designed capacity. 'We're going to have void left to fill,' she says.

Vic and Jamie are passionate about waste in a way that I find infectious. 'It's weird. It's either in you, or it's not,' Vic says. Like Jamie, she thinks of landfill management not as an act of defilement, but one of environmental stewardship; by working here, she is making sure that the waste doesn't end up polluting soil or rivers. 'I feel like I have a responsibility to do this,' she says. It pains her every time she hears about some badly managed landfill creating a stink or causing community protests. Landfills, Vic says, get bad press. Whenever other landfills are in the news, Ellington sees an uptick in complaints. Otherwise, most people forget that they're even here. 'I wouldn't want one on my doorstep,' she says. 'But if I had one on the doorstep, I'd appreciate it if it was run correctly.'

Looking out over the landfill and its rolling hills of dirt, the thing that strikes me is how *clean* it all is. The wind turbines whirr slowly on the horizon, the breeze rustling the treeline that hides the open face from the road. I think back to Ghazipur, where I'd been just a few weeks earlier. You might ask how it is possible that two dumps, one in a global capital city, and another in a tiny rural town in England, can be so grotesquely different, and the answer is of course simple: money. We in the Global North take waste disposal as a given, a basic right, whereas for people who struggle for their basic needs – food, water, education – it's a luxury.

In our debates about recycling and the circular economy, it's worth considering that in low-income countries, 93 per cent of waste ends up dumped. In high-income countries, the figure is only 2 per cent. Yet the world's population is growing fastest in the Global South, where waste management ranges from poor to nonexistent. Often, these are the countries producing the very things that we're throwing away in the first place. Our solutions for the waste crisis cannot just be for the well-off; they must include everyone, wherever they are, rich or poor. Otherwise the Ghazipurs of the world will just keep getting taller.

After we're done at Ellington, Vic, Jamie, and I drive out to one of SUEZ's former landfills in Seghill, a town a few miles south. The Seghill site closed in 2012. In the UK, as in the United States, waste contractors are legally responsible to remediate landfills and maintain the sites for thirty years after closing, including drawing off the leachate and landfill gas. The waste itself will remain intact for decades, and likely centuries after that, the risk of liner leaks only increasing with time. But on the surface, at least, it's peaceful: a quiet hillside, topped by wheat fields. The hedges are thick with blackberry and hawthorn. Recently planted trees – beeches, I think – are starting to bud. The only sign of this land's true purpose are the methane pipes that protrude from the soil every few hundred metres, like tree shoots. 'We get red squirrels here,' Vic says; endangered animals, finding a new home above the waste. Vic likes it here. As one of the remediation projects on the site, she has opened a nature school, where schoolchildren can come to learn skills like beekeeping – as well as about the garbage, at least temporarily entombed below their feet.

2

SAVE SCRAP FOR VICTORY!

'You only see what people want you to see'

Before a cardboard box can be recycled – before it can be shredded, washed, mashed, dyed, pressed, steamed, and rolled until it is finally itself again – it must first, it turns out, be quite unceremoniously probed. I discover this curious fact one autumn day at a paper mill in Kemsley, Kent, looking out over the Thames estuary. I've come to Kemsley because of something that Jamie Smith said, back at Green Recycling: that the roiling mass of Amazon boxes on the conveyors would be sent to Kent and become new boxes within a fortnight. The circular economy made manifest. This I had to see.

Outside the paper mill, a queue of 14-tonne lorries files sluggishly through the front gate towards the weighbridge, like students reluctantly handing in their homework. Like all recycling, paper and card (or 'fibre', to use the industry term) is a commodity, sold by weight. Knowing what you're buying is crucial. Even after it has passed through an MRF, household waste contains multitudes: grease, food scraps, staples, tape, polystyrene peanuts, plastic films. For the recycler, every bale is a special delivery whose contents are unknown. It's not uncommon for unscrupulous waste dealers to take advantage of this fact by weighing bales down with heavier items: bricks, concrete slabs. Once, the mill workers at Kemsley found an entire engine block wrapped in wastepaper. 'Historically, quality control was someone having a look inside,' Jonathan Scott, the plant's technical operations manager, tells me. 'But of course you only see what people want you to see.' Hence the probe.

A mill worker in high-vis peels the curtain sidings of a lorry back to reveal the cargo within: bale upon bale of junk mail, birthday cards, sandwich packets, delivery boxes, tattered magazines. Compacted and tied with plastic strapping, they have the shaggy look of pompoms. Smaller scraps of paper slip free and float to the ground, as if moulting. The probe itself is automated: a robotic arm picks a bale at random, drills a hole, and inserts an infrared sensor. Each shipment is tested for plastic contamination and for moisture content; cellulose soaks up water (being, well, trees) and an overly soggy load can cut into the mill's bottom line. 'If we're paying £150 per tonne for material, we don't want to be paying it for rainwater,' Jonathan says. I murmur something polite in agreement. Jonathan strikes me as a wise figure: bald and aquiline, monkish, with the gentle stoop of an under-watered tulip. He's been at this job for eight years and knows everything there is to know about fibre – or, as they're fond of calling it at Kemsley, 'beige gold'. The probe is a recent innovation, Jonathan explains. An industry first. I get the sense I'm supposed to be impressed.

In truth, I'm in awe. The routine is incessant, the scale immense. Kemsley is the largest paper recycling mill in the UK by volume, and the second largest in Europe. Owned and operated by the multinational packaging giant DS Smith, the plant receives 2,500 tonnes of wastepaper every day. 'Six lorry loads per hour, twenty-four hours a day, seven days a week, fifty-two weeks a year,' Jonathan says, proudly. In return, it churns out more than 820,000 tonnes of new paper and corrugated case material (the precursor to cardboard boxes) every year – nearly a third of all the fibre recycled in the country. 'Box to box in fourteen days,' Jonathan says.

Fibre is a voracious business. Globally, we use an estimated 416 million tonnes of paper and board annually – a figure that has doubled since the 1980s.[1] Of all the trees felled each year by industrial logging, more than a third are used to produce paper and packaging.[2] In Britain and America, fibre makes up nearly a quarter of household waste, the largest share of any material by weight except food. In 2018, Americans threw out 67.4 million tonnes of it.[3] Of all the materials we throw away, it is also among the most circular – in the UK, 80 per cent of paper and cardboard is recycled.

Once the bales have been quality tested, the trucks roll up to the stockyard below us. A forklift unloads the bales and piles them high, like a toddler stacking blocks. Hundreds are already waiting. Another worker in a Volvo excavator tears the bales open and dumps the contents onto two conveyors. 'They're like automated locusts,' Jonathan says, happily. The concrete is so strewn with paper that were a tornado to pass through, I don't think I could tell the difference.

Jonathan leads me into the plant. This mill was first built in 1923 and has been continuously upgraded ever since, giving the architecture a steampunk aesthetic. Pipes jut from old brickwork and tunnel into new. Fibre dust lies thick on every surface, like ash after a forest fire. Inside, conveyors tip the material into giant stainless-steel vats, where it is mixed with immense quantities of water. 'It's a big food blender, essentially,' Jonathan says, loudly. Paper-making is a thirsty process: producing a single kilogram of paper requires up to 170 litres of water.[4] Inside the vats, the pulp is further diluted and any inks washed out, while a rope-like filter called a 'ragger' pulls out impurities (staples, wire, plastic) which curl out of a nearby pipe. 'Once, those would have been landfilled. Now we recover most of it,' Jonathan says. Navigating the maze of tanks and byzantine pipework feels a little like strolling through a herbivore's small intestine.

A cardboard box, Jonathan explains, is not comprised of a single material: 'You've got liner, fluting, and then you've got the five layers of corrugated.' Kemsley's main products are corrugated case material; Kraft top, which is used for higher strength products; and white liner, which has a printable surface on one side, and is used for display packaging. Each runs on a different line, and to a different recipe. White paper must be bleached. To corrugated, a mixture of dyes is added, to achieve a just-so shade of brown.

We reach the wet lines, inside a cavernous skylit hall, where Jonathan advises I put in my ear plugs. Two giant German paper-making machines stretch into the distance, whirring incessantly. The machines spray the watery pulp onto spinning belts, where heaters and vacuum suction machines draw off the liquid until the fibres form a sheet of solid paper again. The heat is monstrous. 'It's the humidity that does it,' Jonathan says, cheerfully. Each machine will spin out 60km of fresh fibre every hour.

At the end of the process, the paper is wound onto 25-tonne rolls, which look like they would flatten a bus. We watch one of the plant workers change the finished spool with a crane. 'It makes one of those every thirty-five minutes,' Jonathan says. It's uncanny, seeing them here stacked on end, the rolls ringed like giant sequoia — only wrongly reassembled, like those chicken fillets in the shape of dinosaurs.

After a few years in the reputational doldrums, fibre is in the midst of a comeback. Recent bans of plastic straws, bags and wrappers have led to a surge in their paper equivalents; Big Plastic's loss has been Big Forestry's gain. The biggest driver has been the growth of online shopping. An estimated 20 billion parcels were shipped in the United States in 2020, almost all of them boxed, wrapped or tucked into an envelope. The demand for fibre is so great that it has lately started to outstrip supply. A surge in online shopping during the pandemic, combined with a shortage of new fibre from abroad, caused cardboard prices in the UK to increase tenfold.[5] The industry was so desperate for material that it put out a plea for people to raid their garages for old boxes.[6] 'You look at North America, Scandinavia. Lots of woodland, cheap energy, make paper from trees,' Jonathan says. 'In the UK? No forest, expensive energy, and 70 million people living in quite close proximity.' He says this as if the solution is obvious. 'The urban forest – harvest that.'

Recycling is as old as thrift. Five hundred thousand years ago, Prehistoric humans repurposed stone axes and tools that were blunted or broken.[7] The first paper, produced in China around AD 105, was made with old rags and plant waste, and continued to be until the nineteenth century.[8] For centuries, books were bound with the scraps of old manuscripts.[9] The Maya, Romans and Greeks reused old tiles and brickwork in new buildings; remnants of this practice, known as *spolia*, are visible throughout the Mediterranean. Change ruler or religion? Cheaper to recarve the head of a statue or swap in a new one than to mine, hew and lug around all that new marble. For almost all of human existence, raw materials were expensive, rare, and precious, and so objects would have many lives. That materials should be recycled was self-evident. Even the word *recycling* did not take on its modern meaning until the environmental movements of the 1970s. Prior to that it was a technical term used to describe the process of recirculating by-products inside oil refineries.

By the Industrial Revolution, recycling was an intrinsic part of modern life. Whenever new factories sprang up, entrepreneurial scrappers found ways to profit from their spoils. Metal waste could be melted back down. Coal waste could be transformed into creosote and naphtha. Dust and ashes could be made into construction materials. Slaughterhouse waste would be rendered into soaps and glues and candles, bones ground into fertiliser. In 1851, Henry Mayhew published *London Labour and the London Poor*, his landmark exposé of the conditions of the Victorian working classes. In it, Mayhew offers a vivid taxonomy of the city's poor and wretched: beggars, pickpockets, knife swallowers, street sellers, sweeps, prostitutes – and waste pickers. The London that Mayhew describes is a complex recycling economy. Rag and bone men pull their carts from house to house, buying up 'waste-paper, hare and rabbit skins, old umbrellas and parasols, bottles and glass, broken metal, rags, dripping, grease, bones, tea-leaves and old clothes.'[10] Mudlarks pick along the Thames shore in long velveteen coats, loading their oversized pockets with waste washed in on the tide, while 'toshers' rove the sewers for scraps of copper. Then there are the 'pure-finders', who pick dog shit to sell to leather-makers.* 'In London,' Mayhew wrote, 'nothing is to be wasted.'[11]

That isn't to valorise Mayhew's account. Most of the waste pickers that he describes live in extreme poverty, 'the cheapest and filthiest locality of London', where 'foul channels, huge

---

* In case you're wondering, as I did: the poop contained enzymes that soften or 'purify' the leather, hence the name.

dust heaps, and a variety of other unsightly objects, occupy every open space, and dabbling among these are crowds of ragged dirty children.'[12] Victorian London's waste problems were well documented, not least by Charles Dickens, who filled books like *Bleak House* and *Oliver Twist* with images of filth. In his last novel, *Our Mutual Friend*, Dickens describes wastepaper which 'hangs on every bush, flutters in every tree, is caught flying by the electric wires, haunts every enclosure'.* The same was true in most major industrialised cities of the time. In New York, so many cartloads of garbage were being dumped into the ocean that it was washing up on the New Jersey shore.[13]

In London, the growing waste crisis was taken up by social reformers, chief among them the barrister and campaigner Edwin Chadwick. Chadwick was an imposing figure, 'utterly lacking in tact, graciousness, and diplomacy', as his biographer Samuel Finer put it.[14] As an administrator for the Poor Laws, Chadwick had seen first-hand the wretched conditions of Britain's slums and workhouses. In 1839 he was commissioned to take a survey of the conditions of the nation's working classes. Chadwick's *Report on the Sanitary Condition of the Labouring Population of Great Britain*, published in 1842, caused a sensation. It described British cities and towns in fetid squalor, 'decomposing refuse close upon their doors and windows, open drains bringing the oozings of pigsties and other filth'.[15]

Chadwick and sanitarians were motivated by the miasma theory, the idea that the diseases and epidemics regularly tearing through the population – cholera, typhus – were caused by vapours, or 'foul airs'. Waste, therefore, was not just fetid and degrading, but killing people. (Although the mechanism was

---

* He's describing paper, but it's eerie that Dickens could quite as easily be describing plastic bags today in much of the world.

wrong, the fundamental idea that unsanitary conditions spread disease was correct.)[16] Something had to be done. In 1848 the British Parliament passed the Public Health Act, the first in a run of measures designed to clean up Britain's cities. But it wasn't until 1875 that an amended version of the law cemented Chadwick's lasting sanitary contribution: it authorised local councils to provide 'receptacles for deposit of rubbish' to be collected weekly. These 'dustbins', and the idea of waste collection as civic infrastructure, would help to transform the city – and eventually, modern life.

Soon, major cities were embarking on sanitation drives, installing sewers, clearing the dust yards and dung heaps. The principle of sanitation as a public good spread like one of the many viruses it was designed to cure. In 1884 Eugène Poubelle, the *préfet* for the Seine – the regional administrator in charge of Paris and its surrounds, then home to over 2 million people – went further than Chadwick, and decreed that every dwelling in the city should be provided with no less than three covered containers to dispose of their waste: one for compost, another for paper and textiles, and another for crockery and shells. The waste itself was collected in the mornings by horse-drawn carriage. The idea was fought both by landlords and the city's waste pickers, who thought it would put them out of business – but the initiative stuck. Even now, the French for bin is *la poubelle*.*

Perhaps my favourite of the sanitarians is Colonel George E. Waring, New York's commissioner in charge of street cleaning from 1895 until 1898. A gregarious, thrice-married Civil

---

* There is a street named after Poubelle in the 16th arrondissement. A great honour for him, but perhaps less so for its inhabitants, who now live on what essentially translates as 'bin road'.

War veteran, Waring rode to work sporting a pith helmet and handlebar moustache, and regaled the public with published accounts of his adventures through Europe and America on horseback.[17] Waring was a polymath, fascinated by the new science of sanitation. Early in his career he had been involved in the creation of Central Park, and after the war worked to construct sewer systems from upstate New York to Massachusetts. In Manhattan, Waring outfitted his new Street Cleaning Department in bright white uniforms and caps; to build public support, he paraded his new 'White Wings' down Fifth Avenue with marching bands. Like Poubelle, Waring mandated that New Yorkers separate their trash into three receptacles: one for food waste, one for dry rubbish and another for ash. The waste was carried across the Hudson by boat to Brooklyn's Barren Island, where Waring created America's first waste-sorting plant – a kind of precursor to the modern MRF – where eight hundred of the city's most desperately poor workers, mostly immigrants, picked through the trash for recyclables.[18] Waring's innovations were unfortunately short-lived. He contracted yellow fever and died in 1898, and his successors in City Hall abandoned many of his schemes. But his idea for city-wide recycling remained, and when the environmental movement resuscitated the idea of municipal recycling schemes in the 1970s, this time it stuck.

Today, the idea of recycling as a public good is commonplace, if not universal. Around the side of my house, behind a trellis climbed by white roses, are no fewer than seven bins. There is a bin for rubbish (purple); another for mixed recycling (black); one for paper (blue); one for food waste (dark brown); one for garden waste (light brown); and two of our own, for compost. They're drab, perfunctory things, each a little grimy in its own way, and in at least one case heavily cracked from being tossed around on collection day. I have always been a decent recycler – B+, could do better – but the more I have learned about waste, the more fastidious I have become. Who knows how many hours I have spent washing out yogurt pots, preparing plastic bottles (the trick is you crush them, then put the lids back on), tearing oily shapes from pizza boxes so as to salvage the rest. Hertfordshire, the county in England where I live, employs what is known as a multi-stream collection* system: by separating my source material into all those bins, I am doing the work of an MRF line picker in advance, and therefore making recycling simpler (and waste companies' profits a little easier to come by).

Why do I do it? Muscle memory, partly. In Britain, we now spend our lives being taught that recycling is a civic responsibility, and are constantly harried into it by ad campaigns, signs and labels. Partly, I do it to maintain my own sense of optimism, every mindful action – an extra second taken to rip the plastic window out of a sandwich box – a small gesture of care. Mostly I do it to assuage my guilt about the things I throw away. *This is for the planet*, I tell myself. *I'm doing something good.*

And recycling *is* doing good, mostly. The environmental benefits of recycling are manifold. Recycling an aluminium can requires roughly 92 per cent less energy and emits 90 per cent less carbon than making one from virgin metal; for every ton of aluminium saved, you're also saving eight tonnes worth of bauxite ore from being mined from the ground.[19] Recycling one ton of steel requires just a quarter of the energy of mining it new, cuts the associated air pollution by 86 per cent, and

---

* As opposed to single stream, in which recycling is all collected in one bin. Multi-stream approaches vary widely, depending on where you live, but the principle – a cleaner waste stream for easier recycling – is the same.

saves around 3.6 barrels of oil.[20] Recycled glass requires 30 per cent less energy to produce, paper 40 per cent less,[21] copper 85 per cent. By recycling most materials, we're not only reducing the greenhouse gases required in production, but the environmental damage caused by extraction: the logging, mining, processing, and transportation required in replacing that item with new. Recycling creates less water and air pollution. It reduces the amount of waste that then ends up in landfills, dumped, or burned.

It is also better for the economy. Globally, the recycling industry employs millions of people; the market for scrap metal alone is worth more than $280 billion.[22] Studies have shown that recycling schemes create 70 jobs for every one that would be created by landfill or incinerators.[23] And the scale is enormous: 630 million tonnes of steel scrap is recycled globally every year.[24] It's estimated that 99 per cent of the metal in scrapped cars, for example, ends up reused. Of all the copper ever mined, 80 per cent is still in circulation.[25] In the UK, three-quarters of glass waste is recycled into new bottles, fibreglass or other materials. There is little doubt that recycling most materials is a better solution than burying or incinerating them.

The problem, as you have probably guessed by now, is plastic.

Plastics don't just end up as waste, they begin as waste. The chemical building blocks of many plastics – ethylene, benzene, phenol, propylene, acrylonitrile – are inherently waste products, created in fossil fuel production.[26] Most were habitually burnt or released into the atmosphere until the 1920s and '30s, when enterprising oil and coal industry scientists discovered ways to solidify them into polymers. Carbon, captured.

Although plastics existed prior to the Second World War

(the British inventor Alexander Parkes patented the first manmade plastic, Parkesine, in 1862), it was the war that triggered the plastics revolution. Parachutes, cockpits, radar systems, combat boots, ponchos, bazooka barrels – all were moulded with the new materials being churned out of the labs at the likes of Dow Chemical and DuPont. From 1939 to 1945, global plastic production increased nearly fourfold.[27] Soon, many of those same materials – Nylon, Perspex, polyethylene film – were flooding into households, newly empowered with cheap credit; plastics bought on plastic.

For all their flaws, plastics are a miraculous class of materials: cheap, elastic, lightweight, durable, capable of taking on almost any shape or colour. Suddenly, here were materials that could be stretched into sandwich wrap or hardened into a car bumper, moulded into an intravenous tube or spun into a T-shirt. Ivory, wood, cotton, whalebone, leather – plastics could imitate them all. Pretty much anything could be plasticised. Pretty soon everything was.

In the post-war boom, the burgeoning middle classes were buried under an avalanche of stuff. *Disposable* stuff. 'The future of plastics is in the trash can,' Lloyd Stouffer, the editor of *Modern Packaging* magazine, declared at a Society of the Plastics Industry conference in 1956.[28] Early plastic products had been designed to be high quality, to take advantage of their innate permanence. But Stouffer urged his colleagues to abandon durability and 'concentrate on single use'. In truth, they didn't need telling. Empowered by a glut of cheap raw materials, packaging companies had already begun pumping out single-use versions of almost anything you could buy: pots and pans, plates, cutlery, towels, dog bowls. Drinks once sold as refills were now packaged in one-way bottles or plastic cups, designed to be tossed after one use.[29] There were disposable curtains (to make it easier to follow the latest decor trends) and even, for a time, paper dresses.[30] *Life* magazine celebrated the arrival of 'throwaway living', which would, it was said, liberate housewives from their daily drudgery and eliminate tedious chores. Impermanence was in vogue.

Five years later, Stouffer was addressing the SPI conference again, congratulating the plastics industry for, in his words, 'filling the trash cans, the rubbish dumps and the incinerators with literally billions of plastic bottles, plastic jugs, plastic tubes, blisters and skin packs, plastic bags and films and sheet packages'. (To Stouffer, this was a good thing.) Inevitably, all of that disposable stuff had to go somewhere, and so soon waste collection was overwhelmed. In the United States, this new influx of garbage filled gutters and verges. In 1962, the writer John Steinbeck published *Travels with Charley*, in which he described driving across America, cities 'like badger holes, ringed with trash – all of them – surrounded by piles of wrecked and rusting automobiles, and almost smothered with rubbish'. Industrial pollution was rife, although the Environmental Protection Agency would not be formed for another decade. Trash was out of control.

The very characteristics that make plastics extraordinary are those that make them so problematic as waste. The covalent bonds that bind them together are, at a molecular level, durable to the point of stubbornness. Almost nothing alive digests them, so they do not decompose.[31] Like oil, they are slippery, elusive. The heaviest plastics sink without a trace, while the lightest film can catch a breeze and travel continents. A few years ago, I travelled to the South Pacific on a research expedition with the submarine explorer Victor Vescovo as he attempted to reach the deepest point in each of the world's oceans. There, 10,900m below the surface, in the lightless void of the Mariana Trench, he looked out of the window and saw a plastic bag.

When plastics *are* broken down, by ultraviolet radiation, by the elements, or by force, they do not disintegrate so much as divide, their chain-like structures splitting into smaller and smaller pieces of themselves. Macroplastics become microplastics become nanoplastics. By then they are small enough to enter our bloodstreams, our brains, the placentas of unborn children. The impacts of these materials on our bodies are only just beginning to be understood; none are likely to be good. There are no firm figures on how many micro- and nanoplastic particles now pervade our planet, although one recent study estimated that the upper ocean alone contains 24.4 trillion microplastic particles.[32] Polymers, from the Greek: *many parts*.

On a Thursday in March I arrive in Seaham, just south of Newcastle on England's north-east coast — not too far, in fact, from Ellington landfill. I've come to visit Biffa Polymers, a gleaming new plastics recycling plant perched on a windswept hilltop overlooking the North Sea. This stretch of England has known heavy industry for centuries: coal mines, foundries, brickworks, bottleworks. At the height of the Industrial Revolution, ships left port at Seaham loaded with the materials to build the British Empire. The scars of that industrial legacy are still written into the cliffs and rocky sands. The recycling plant overlooks Chemical Beach, so named for the nineteenth-century chemical works that operated here, turning out soda crystals and magnesia; its wastes still wash up on the beach as sea glass even now. Further along the coast, at Blast Beach, the old coal tips that remain have eroded to reveal cliffs of jet-black slag.

The Biffa plant itself is a sterile grey-and-blue box, numbingly corporate. I'm met by Chris Hanlon, Biffa Polymers' commercial director, an amiable company man with a grey goatee and heavy bags under his eyes. Chris offers me a coffee, but he looks like he needs it more than I do. Upstairs, the administrative suite is airy and bland, and decorated with what appear to be stock photos of beaches – all of them, I notice, idyllic and plastic-free. In fact, the only sign of what really happens here is the graphic laser-cut into office partitions and stencilled on the stairwell walls: the outline of the molecular structure of polyethylene terephthalate, otherwise known as PET.

Even by plastics' standards, PET (pronounced as one word, like a puppy) is an extraordinary material. First discovered by DuPont scientists in 1941, it can be transparent or coloured, set into firm objects, flexible containers, or spun into fibres. Like HDPE and polypropylene, PET is thermoplastic – that is, it can be easily melted down and recycled (as opposed to ther-mo*set* plastics, such as polyurethane, which cannot). But its real success lies in what plastics chemists call its intrinsic viscosity, or IV. 'It's basically a measure of the density of the polymer,' Chris explains. In 1973, another DuPont scientist, Nathaniel Wyeth, realised that he could take advantage of PET's natural flexibility to hold carbonated beverages. 'A carbonated Pepsi or Coca-Cola bottle is a small pressurised container. The packaging has to be able to hold the lid on effectively, so that when the gases build up, it doesn't explode,' Chris says. 'You need something with a quite high IV – you need PET.' This facility, opened in 2018 at a cost of £27.5 million, is designed to recycle up to 57,000 tonnes of PET every year – the equivalent of 1.3 billion bottles.[33] (That might seem a lot, but 480 billion plastic bottles were sold worldwide in 2016.)[34]

Before joining Biffa, Chris worked in hazardous waste treatment. 'I got tired of landfilling and burning things,' he says.

These days he is at home among the pipework and lab benches that come with recycling – the alchemical strangeness of transmuting matter back into itself. When he started at Biffa in 2004, the only real market for recycled plastic was dairy. In the UK, 90 per cent of fresh milk is sold in standardised HDPE bottles, of which 79 per cent are recycled.[35] 'We've been recycling milk bottles back into milk bottles since 2007,' Chris says. Just like in Delhi, plastics people speak in code: the whitish, translucent HDPE used for milk bottles is 'HD Nat', the richly dyed variety used in fabric softener bottles, 'Jazz'. 'Don't ask me why,' Chris says. 'We just do.'

The truth is that a few years ago, Chris considered quitting the recycling business. 'I actually got quite disillusioned, because it didn't feel like we were going anywhere,' he says. In those days, plastics recycling was a niche industry, barely profitable. 'Then *Blue Planet* happened, and things just went ridiculous.' In 2017, the BBC released the documentary series *Blue Planet II*. The show culminated in a vivid episode about plastic pollution that depicted marine life choking in six-pack rings, and triggered a moral awakening over plastics in Britain. '*Blue Planet* changed the game. It started off as a horror story about plastics, but really did our industry a lot of good,' Chris says. 'It really focused people's attention on plastics and the circular economy.' Packaging brands have been falling over themselves to adopt recycled plastics ever since.

For Biffa, plastics recycling is an inevitable extension of its waste collection business. The PET is collected through Biffa's MRFs. 'They segregate out the PET and HD Nat, we reprocess it back to a pellet, and then we sell it back to the drinks manufacturers,' Chris says. 'When we started in HDPE, there was no way we were selling our content for anything less than a discount on virgin [material]. We actually had a model that said, "The virgin price this month is this – our price will be 20 per cent lower." That was the selling point: it was cheaper,' he says. Today, due to soaring demand from companies trying to rescue their environmental image, the price of recycled PET is £300–500 per tonne higher than virgin. 'Biffa Polymers turns over £200 million a year. When I started, it was £2 million,' Chris says. 'It's incredibly profitable.'

He loans me a fluorescent yellow vest and a helmet and leads me downstairs onto the factory floor. Just like at DS Smith, the loading bay is crowded with bales, each holding half a tonne of PET. The bottles strain against the baling wire, as if unlawfully held. Whenever I've seen waste pickers' PET hauls they're always iridescent and strangely beautiful – but here, with the air crushed out of them, they're oddly lifeless. The coloured labels and lids stand out among the grime like flecked wool.

'These bales look pretty dirty,' I say, as we wander the stacks.

'The quality in the UK at the moment is not as good as you would think,' Chris says. This is partly a result of how we collect recycling in this country, with many materials – card, plastics, glass – thrown in together. That makes sorting a nightmare. That problem is made worse by 'wishcycling', people putting things into the recycling that they think are, or ought to be, recyclable, but in fact are not. 'If you look at European-quality PET in France or Germany that has been collected through a DRS [Deposit Return Scheme], the material is beautiful,' Chris says. 'It looks like it's hand-picked.' Ensuring a pure stream of waste materials is the single biggest priority of any recycling facility. He pulls a rogue aluminium can from a plastics bale. 'We'll recover that,' he says, holding on to it. Even with all this mechanisation, some things are better done by hand. 'It's very visual,' he says. 'We have no other way of doing it.'

* * *

The word *plastic* is misleading. Plastics are not one material or even one chemical but tens of thousands, which vary wildly in form, toxicity, and function.* (The word plastic refers only to their common feature, plasticity.) The confusion lies in their similarity – while anyone can tell a PET bottle from a polystyrene peanut, far fewer can easily recognise the subtle differences of thickness between linear low-density polyethylene and polystyrene film. As Akbar told me in Delhi, it would take a year to learn to recognise different plastic categories the way they do, by sight, touch, and memory.

The way the waste industry has historically got around this problem is by using labels. It is not as simple as it sounds. In the UK, *Recycle Now* lists twenty-eight different recycling labels that can appear on packaging. There is the Mobius loop (three twisted arrows), which indicates a product can technically be recycled, but not that it will be. There is the Green Dot (two arrows embracing), which indicates nothing about where it will end up, merely that the producer has contributed money to a European recycling fund. There are labels that say 'Widely Recycled' (i.e. not everywhere recycles it) and 'Check Local Recycling' (i.e. almost nobody does). The most ubiquitous symbol, the one you'll recognise, is the International Resin Identification Code: three 'chasing arrows' with a number at its centre.[36] The symbol was first created in 1970 by Gary Anderson, a student at the University of California, who won a competition held by the Container Corporation of America to design a recycling logo.[37] It was not until 1988 that the Society for the Plastics Industry co-opted the symbol and added the number at its centre, as a means to help waste processors identify which plastic a product is made out of.[38]

* Which is why I've used the plural.

The resin code is a mess. Everyone I've met who works in the waste industry abhors it. The symbols are useless to waste processors (who don't have time to check the thousands of labels that roll past on a conveyor every minute) and confusing for consumers, most of whom mistakenly assume that anything with the numerical code on will be recycled, even though generally only plastics numbers 1–4 are, if you're lucky. This is not entirely by accident. As reported by *PBS Frontline*,[39] the plastics industry lobbied hard for US states and other countries to adopt the resin code, despite knowing since the early 1990s that it confuses consumers.[40] Numerous studies have shown that existing recycling labels mostly serve to trick consumers into thinking that products are recyclable, even when they're not.[41] This, in turn, increases 'wishcycling', and thus contaminates things that *are* actually recyclable.[42] One thing that recycling *does* do is assuage consumers' guilt about their waste. If a product is seen as recycled, or recyclable, it makes us feel better about buying it; we have little way of knowing whether the claims on the labels are true, or not.

Of course, just because something is recyclable, does not mean it will be recycled. Take, for example, perhaps our most famous story about recycling: the scrap drives of the Second World War. At the height of the conflict, governments on both sides urged their citizens to save scrap materials for the war effort. All over Britain and America, posters and pamphlets exhorted the public to 'Save Scrap for Victory!' and 'Throw Your Scrap into the Fight!' In Britain, towns tore down iron fences for scrap; the London Underground donated its safety railings. Everything had its worth: a bucket, one US government pamphlet declared, contained enough metal for three bayonets, while the aluminium inside a washing machine would make four incendiary bombs.[43]

The wartime scrap drives are legendary; in Britain, you can still buy the posters in museum gift shops. Propaganda, however, rarely resembles reality. The truth is that much of the material collected during the wars was never used. It was simply impractical to collect, sort and recycle the hodgepodge of donations with a war still going on. (In fact, some historians argue that the main impact of the scrap drives was to actually *crash* the market for recycled materials.)[44] In the end, much of what was collected was landfilled or abandoned. It's said so much metal was dumped into the Thames that it confounded ships' compasses. But the truth about the scrap drives was overlooked, in case reality were to dent the public's morale, and with it its long-term appetite for recycling. The system is built on trust; without trust, the system might collapse.

We reach the lines. Ordinarily there would be a dozen guys working this shift, moving product. But at the moment the lines are mercifully paused, and so sit quietly. We navigate the carefully engineered weave of conveyors and pipework. There's dust on the machinery, light, like lint. 'It's PET fluff. It gets everywhere,' Chris says. As at most recycling facilities, the PET bales must be re-sorted before processing, to reduce contamination as much as possible. The waste is fed upwards onto a red conveyor, and carried through yet another Rube Goldbergian system of sorting machines. Labels are typically made from polypropylene rather than PET, so must be removed in a specialist de-labeller. Things like Lucozade, which have a shrink–wrap label around the entire bottle, are 'a pain in the ass,' Chris says. Caps are also a different material, and so are removed from the line and recycled separately. 'We do 5 or 6 tonnes of caps per day.' Any foreign materials are extracted and bagged off below the line, at ground level; Chris tosses the can he's been cradling into a bag marked for metals. The conveyor eventually snakes up and into a drab grey tower which houses a manual picking station. Inside, plant workers remove by hand anything that gets through the mechanical screening. 'They're looking for anything that the process might have missed. It might be rags, it might be Lucozade bottles, it might be dead cats, or engine blocks,' Chris says. (Seriously, where are all these engine blocks coming from?)

The bottles are then fed into a granulator, which chops the PET into flakes. The flakes are washed in hot caustic soda, and passed through a sink-float system. 'PET sinks, HDPE and polypropylene will float,' Chris explains. At the end of the line, Chris shows me some fresh flake, poured into white transport bags the size of chest freezers. I realise I'd seen it before, mixed into the daily cover at Ellington landfill.

I run my hand through the flake. It is undeniably beautiful: jagged and polychromatic, the colours of oil on water. Most oil deposits are thought to have been laid down between 10 and 180 million years ago, when dead zooplankton and algae sank to the seafloor and were buried in layers of sediment, before being slowly transformed into hydrocarbons by heat and pressure and time. Matter perhaps as old as the dinosaurs, falling through my fingers like confetti.

In 1953, with a backlash against America's garbage crisis growing, the US state of Vermont passed a law banning the sale of non-refillable bottles. It was the opening salvo in the fightback against disposability, launched amid mounting pressure on the packaging industry to clean up after itself.

The industry moved quickly. That year, it helped to establish Keep America Beautiful, an environmental non-profit organisation whose founders quietly included the American Can Company, the Owens Illinois Glass Company (a disposable bottle manufacturer), PespiCo, Coca-Cola, and the makers of the Dixie cup.[45] Keep America Beautiful quickly became one of the loudest voices in the fight against America's waste problem, launching ad campaigns and cleanup drives across the country. But the group did not place the blame for the waste problem at the feet of the companies producing and promoting single-use products. Instead, as the journalist Heather Rogers recounts in her book *Gone Tomorrow*, it pointed its well-funded fingers at a newly-coined enemy: 'litterbugs'. Garbage, according to Keep America Beautiful, was the responsibility not of companies, but *individuals* – culminating in 1971 with its notorious 'Crying Indian' advert, which declared: 'People start pollution. People can stop it.'

The tactic spread. All over the world, the packaging industry invested in campaigns which reframed waste as 'litter'. In the UK, the charity Keep Britain Tidy – whose Tidyman logo, a stick figure throwing paper into a trash can, adorns most packaging 'to remind people to dispose of their waste responsibly'– has long been funded by donations from the packaging industry, including Coca-Cola, Nestlé, and McDonald's. (Its first chairman, according to the historian Frank Trentmann, was a managing director at an oil company.)[46] Those same corporations, meanwhile, have consistently lobbied against legislation designed to make packaging producers take greater responsibility for waste pollution. This tactic – to deflect blame onto consumers – is a longstanding corporate strategy: the carbon footprint, for example, was popularised in the early 2000s by the oil giant BP.[47] We're told to 'gamble responsibly' and 'drink aware'. The word *less* never comes into it.

Litter, however, was only one piece of the packaging industry's defence campaign. The other was recycling. Beginning in the 1980s, Big Plastic began to spend tens of millions of dollars promoting recycling as the solution for our waste problem, often through industry front groups with names like the Council for Solid Waste Solutions, the American Chemistry Council, and the Clean Europe Network.[48] The primary purpose for this big recycling push was to avoid legislation – at the time, environmental campaigners were calling for a ban on polystyrene, then used in fast food packaging. 'No doubt about it, legislation is the single most important reason why we are looking at recycling,' Wayne Pearson, then executive director of the Plastics Recycling Foundation, an industry consortium that included DuPont, Coca-Cola and PepsiCo, told the *New York Times*.[49]

The truth is that, in private, even the plastics industry did not consider recycling a realistic solution to its waste crisis. Internal studies cited plastic's degradation problems, as well as the challenges with sorting, and with separating mixed materials. 'There is serious doubt that [recycling plastic] can ever be made viable on an economic basis,' an industry insider told a 1974 SPI meeting.[50] (While that's not necessarily true for PET today, it remains true for many plastics.) In the decades since, whenever its profits have come under pressure – whether by packaging taxes, deposit return schemes (so-called 'bottle bills') or plastic bans – the plastics industry's solution has been to push more recycling. As Larry Thomas, the ex-president of the Society of the Plastic Industry, confessed to NPR: 'If the public thinks that recycling is working, then they are not going to be as concerned about the environment.'[51]

Over the years a kind of playbook emerged: plastics companies would make big promises about moving to more recycled content and even open new recycling facilities, only to abandon them when attention moved on. Coca-Cola, for example – which sells 3 million tonnes of plastic, or the equivalent of 108 billion 500ml bottles per year – recently pledged to make all of its 'on-the-go' bottles in the UK from 100 per cent recycled material. At the same time, it has lobbied *against* proposals to reform solid waste collection and reduce plastic pollution, such as the introduction of deposit and return schemes. In a 2016 internal document leaked to Greenpeace, the company grouped forthcoming legislation under the headings 'prepare', 'monitor', and 'fight back'.[52] Proposals in the latter categories included 'refillable quotas' as well as 'increased collection and recycling targets'. (An annual survey conducted by Break Free From Plastic found Coca-Cola to be the worst plastic polluting company in the world, a title it has held for four years running.)*[53]

This isn't without precedent. In the early 1990s, Coca-Cola announced a goal to make its bottles from 25 per cent recycled plastic, only to abandon the target four years later once consumer and political pressure had lifted.[54] In 2007[55], the company made headlines again when it set out to 'recycle or reuse 100% of its plastic bottles in the US' and to achieve this, opened the 'world's biggest PET recycling plant' in Spartanburg, South Carolina. In reality, the company missed its recycling target, and quietly shut down the plant two years later. Coke's target of using 10 per cent recycled plastic in its bottles by 2010? Missed. It set a target of 25 per cent recycled content in its bottles by 2015, and failed to hit even *half* that. They're not alone: PepsiCo and Nestlé, among others, have all previously failed to reach plastics recycling targets.[56] This is

---

* In case you're wondering, the rest of the top ten were: PepsiCo, Unilever, Nestlé, Procter & Gamble, Mondelēz, Philip Morris, Danone, Mars, and Colgate-Palmolive.

partly a failure of journalism: pledges get news coverage. Few ever check later to see if they come true.

The plastics industry's tactics can be even more duplicitous. For example, the Ban the Bag project, launched by the American Recyclable Plastic Bag Alliance, is actually an industry front group dedicated to *prevent* the passage of plastic bag bans.[57] One of its methods is to lobby for the passage of 'pre-emption' bills – laws which ban bans of plastic bags.[58] (Still following?) In the US, twenty states have passed versions of these laws, many with almost identical wording. It turns out the lobbyists sent out a form.[59]

Chris leads us into another part of the plant, which smells oddly like burning waffles. Here the flakes are fed into a Starlinger extruder, a tube-like and inevitably German machine with pipes protruding from it like cephalopod limbs. The machine heats the plastic to over 250 degrees, melting it back down into a liquid state. While in its liquid form, Chris's engineers add what's known as 'masterbatch'. Masterbatch is another one of those plastics industry euphemisms, in that it doesn't actually refer to a single chemical, but is rather a catch-all term for the chemical mixture added to a plastic during the manufacturing and recycling process. More than 10,000 additives can be used to make plastics, of which around 2,400 are potentially hazardous, according to EU safety standards, including plasticisers, flame retardants, dyes, lubricants, antistatic compounds, deodorisers, and foaming agents. The exact recipe depends on the base plastic being used, and the purpose of the end product. The plastics industry is notoriously secretive about these additives; a recent study found that more than 2,000 known plastics additives have been 'hardly studied' for their impacts on human health and are under-regulated in many parts of the world.[60]

'The only thing we add here is some dye – we add some black masterbatch to it, to make it a better colour,' Chris says, slightly defensively. 'There's nothing in terms of other chemicals or other plastics.' Once it's melted, the liquid PET is forced through 5mm holes, until it solidifies as long, spaghetti-like strands, which are chopped up into plastic pellets. It's these pellets, known in the industry as 'nurdles', that will be shipped off to clients to be blown back into new bottles, or other PET products.

Near the extrusion machine, I see a wheelbarrow containing a contorted lump of glossy grey plastic. It has the strange, stringy texture of soft candy. Chris explains that it is 'filter purge' – PET mixture that has backed up in the machine and needs systematically cleaning out. 'It comes off boiling hot and can stay warm for days on end,' he says. There's a larger container full of it at the back of the factory: a shining, eerie mass. Looking at it, I'm reminded of the story of a group of oceanographers who, in 2006, discovered marine debris washed up on the beach of Hawaii, bound together by bright seams of molten plastic.[61] To describe this new material, the scientists coined the term 'plastiglomerate'. The plastic was likely trash that had been burned on camp fires, and coalesced around the matter on which it was dumped. The researchers speculated that plastiglomerate will likely one day enter into the geological record as a marker of the Anthropocene. Waste as time capsule. *We were here.*

At the final step in the process, the newly recycled nurdles are cooked in two large silos for eight hours. 'That bakes out any volatiles,' Chris says. Underneath the silos, the nurdles are collected in sacks. Unlike the rainbow-like flake they're flat and grey, the colour of concrete. Nurdles have been described as the 'worst toxic waste you've probably never heard of'.[62]

In May 2021, the cargo ship *X-Press Pearl* caught fire and sank off the coast of Sri Lanka, spilling an estimated 1,680 tonnes of nurdles – billions, if not trillions of pellets – into the sea, and triggering an environmental disaster. An oil spill, solidified. As I write this, nearly a year later, it's still being cleaned up.

Chris and I walk up a short flight of stairs to the lab, housed in a small prefab room just above the factory floor. The room is pristine, lab benches crowded with spectrometers and other testing machines. Andy, one of the lab technicians, pulls out a box full of smoky-looking disks of PET. This is what recyclers call 'haze' – crystallisation within the polymer caused by microscopic contaminants and the repeated cycles of heating and melting. 'If it's high haze, it's very smoky,' Chris explains. 'When you blow a bottle [from PET] that's got a high haze, it looks like marble.'

PET, like all plastics, degrades as it is recycled. Every time a thermoplastic is melted down and reset, the polymer chains within it shorten. Volatile chemicals and contaminants accumulate; the polymer essentially becomes corrupted. As a result, most plastics, including PET, can only be recycled a handful of times, and others only once, before they must be either disposed of or locked up in some more permanent state of matter. The result is what is known as downcycling:* plastic packaging is recycled not into new packaging, but into materials like drainpipes, toilet seats, carpet, or artificial turf. The same is true of other materials. Wastepaper can typically only be recycled between three and nine times, as the fibres shorten with each pass through the system.[63] Recycling mills such as DS Smith's rely on a continuous injection of new fibre

---

* Generally speaking, 'downcycling' is used when the end product is lower quality or value than the original.

into the manufacturing stream from new paper and cardboard. Similarly, the PET passing through the line at Seaham relies on a certain proportion of virgin bottles within the bales.[64] For all the talk of the 'circular economy', recycling relies on our ravenous consumption of the new.

As well as haze, the pellets are tested for contaminants: Bisphenol-A (BPA), benzene, formaldehyde. 'All the nasties,' Chris says. This is a legal requirement, for good reason. In 2021, scientists at Brunel University found that recycled PET leached 150 different chemicals into drinks – including toxicants such as antimony, BPA, and numerous endocrine-disrupting chemicals (EDCs) such as phthalates – at a greater rate than bottles made from virgin PET.[65] Like all PET bottles, they also shed microplastics.[66] The health impacts of this are, as yet, unclear, however human and animal studies have shown compelling links between phthalates, a common class of plasticisers, and lower fertility, developmental issues, obesity and cancer.[67] 'I'm 100 per cent happy that the material is safe,' Chris says.

Most of Biffa's clients will have their own standards for haze and contaminant level. 'BPA, if it's above a certain level: no chance,' Chris says.[68] 'Whereas colour is just subjective. It's more of a marketing issue. Coca-Cola might say, "OK, we can accept this if we need it", as it's holding essentially brown liquid.' Haze is one of the challenges of dealing with recycled plastics, particularly those harvested from household waste: every time the plastic passes through the 'loop' it picks up more contaminants. In his office, Chris pulls out a cardboard box, to show me some sample PET bottles with varying levels of recycled content: 10 per cent, 30 per cent, 80 per cent. In each case the haze gets darker, the plastic more opaque; recycling in shades of grey. 'It's going to get harder and harder to keep the clarity, as more recycled plastic flows into the system,' he says. 'The day might come where you get a Coca-Cola in a black bottle.'

Degradation is not the biggest problem with plastic recycling. The biggest problem with plastic recycling is this: we don't really know how much of it actually happens. Take the official national recycling rate, for example. In 2020/21, the official figure for England's recycling rate was 43.8 per cent of household waste, a slight fall on the previous year.[69] However, that figure does not actually measure what is being recycled, but rather *the amount of waste that enters recycling facilities*,[70] a definition hidden inside obscure government documents.[71] This Seaham plant, for example, is designed to receive 57,000 tonnes of PET per annum, but the total *output* is currently, Chris says, around 30,000 tonnes. 'Yield is a problem for us,' he admits. 'At the moment, we're in the mid-fifties.' That is, nearly half of the waste that arrives at the plant is not recycled into new PET. This is largely due to the contamination of the PET that comes from household bins. The remaining waste is diverted to other disposal: metals are sold for recycling, but the non-recoverable material – labels, other plastic – is sent to be incinerated. The yield loss is lower at other plants, Chris says, where the feedstock is higher quality. 'In HDPE, 80 per cent of the HD Nat is recovered, and of the 20 per cent that isn't recovered there's about 12 per cent that is still recycled into polymers, just not food grade.' But in other waste streams, such as the recycling of plastic films, yield can be even lower.

This reality is not limited to plastic. Paper mills and other material processes also suffer losses during the recycling process. Which is to say the nation's top-level recycling figure is a measure not of outcome, but intent: at best a flawed method of data collection, and at worst intentionally misleading.[72] This fundamental data gap is also true elsewhere: Germany, a country hailed by the World Economic Forum for being the best recyclers in the world,[73] employs a similar calculation method, which independent experts have claimed is misleading. In 2017, the environmental consultancy Eunomia calculated that many countries were and are overstating their recycling rates significantly – in Germany's case by 10 per cent, and in the case of Singapore, by as much as 27 per cent. A recent study by The Last Beach Cleanup and Beyond Plastics estimated that the true recycling rate for post-consumer plastic in the United States was just 5 per cent, nearly 40 per cent lower than Environmental Protection Agency estimates.[74] The lion's share of that recycling was just two plastics: PET and HDPE. For several other types of plastics, such as plastic film, recycling is so low as to be virtually nonexistent.

Though some other plastics can be recycled, the reality is that collecting, sorting and separating them is expensive and impractical. Plastic bags and films can be recycled, but companies have historically avoided them because they snag and clog up machinery. Polystyrene can be recycled, but degrades quickly. That's before you get to multi-layered plastic, or MLP, the highly engineered films and tube packaging used to make everything from toothpaste tubes to crisp packets.

The plastics industry's latest answer to this problem is to push 'chemical recycling', which largely uses pyrolysis – burning the plastics in the absence of oxygen – and complex chemistry to reduce plastics down to their constituent parts, which could be used as fuels or new plastic additives. However, there are as yet no companies doing chemical recycling at scale, and several high-profile attempts to do so have ended in failure. In 2017, Unilever launched a much-vaunted chemical recycling plant in Indonesia to process some of the billions of tiny plastic sachets the company has pumped into the Global South, but the plant shut within two years due to financial and technical challenges.[75] Of thirty chemical recycling projects reviewed by Reuters in 2021, half had been either delayed or shut down completely.[76]

Chris, despite everything, is still a defender of plastics. 'The entire problem with plastics is just infrastructure,' he says, as we head back upstairs. 'It's perception.' This is a common industry trope: that plastics recycling would be solved if only there was a supply of perfectly sorted, pristine material. Even if it were possible, it ignores the dirty truth that some plastics are simply not recyclable. 'There's certainly good and bad plastics,' he concedes. 'Let's engineer out the bad plastics. Polystyrene! Expanded polystyrene is an absolute nightmare. But there's alternatives. Standard polypropylene is very recyclable.'

Over the years even Chris has grown sceptical of brands making sustainability pledges. 'They're all talking about reusable, refillable,' he says. 'But in reality, you know they want to shift individual units in plastic bottles.' The recent trend for returning plastics bags and films to in-store recycling points, for example. 'Absolute bollocks, isn't it? It's greenwashing. What they've done is they've sent it somewhere and they've hand-picked it to death to find a little bit of material they can recover, and most of it has gone to landfill or waste-to-energy.' (He's right. A few weeks after our conversation, a *Bloomberg* investigation found that plastic films being returned to Tesco stores by customers for recycling were instead being shipped to Poland and burned inside cement kilns.)[77]

If we really want to fix plastics recycling, Chris says, we should scrap the labels and the International Resin Code. 'That one to seven list shouldn't exist. There should probably be three or four – so get rid of everything else. Get rid of multi-layered films, multi-material packaging. Why are we still selling sand-wiches in cardboard and plastic? Do it in cardboard *or* do it in plastic.' The trick would be to make packaging standardised, like milk bottles. 'A milk bottle is a perfect example of a circular product,' he says. But to really fix plastics recycling for *all* plastics would require a far greater commitment, and investment both at the local and global scale. More than that: it would require everyone within the recycling industry to tell the truth about what is being recycled, and what isn't.

I wish I could have Chris's faith in plastics. But for weeks after I get home I think about what he said about yield, as I tirelessly wipe out yogurt pots and tear films off PET packag-ing. The reality is that, while recycling works (and for metals, glass, and to a lesser extent fibre, it *does* work), there is a lot of waste that is not ending up recycled at places like Seaham. For those materials, the problem plastics that Chris talked of, my placing them in the recycling bin was just the start of a far longer journey.

3

THE WORLD'S GARBAGE CAN

'The plastic keeps coming in'

The stories began in early 2018. Always different, always the same. There would be a village, a town. Then, suddenly, plastic factories would arrive. Small operations: shredders, extruders. Nurdle factories. Men would appear, advertising work, profit to be made. Then: smoke. The smell of burning polymers in the night. Locals reporting headaches, breathing problems. In one case, children started to develop skin rashes.[1] Dumps accruing in the fields, in waterways, in ditches, the labels invariably in foreign tongues.

Perak, Malaysia: *Heinz beans wrappers,[2] Flora margarine tub. Yeo Valley yogurt pot. Listerine mouthwash. Recycling bags from four London councils. Cat food pouches from Spain. Wafer packets from Germany. Factory offcuts from Australia. US postal service boxes. Fiji water bottles, Made In USA.*

East Java, Indonesia: *Gatorade bottle. Arizona Sweet Tea. Capri Sun packets from the UK. Dish soap from Finland. Whiskas cat food, Hershey's cookies. Even the occasional banknote, slipped accidentally into the trash: US dollars, Canadian dollars, Russian rubles, Saudi riyals, Korean won.*

Adana, Turkey: *Andrex toilet roll wrappers. KP salted peanuts packet, Asda cashews. Marks & Spencer bacon. McCain's home fries. Cherry Pepsi Max. A Tesco plastic bag, emblazoned with flowers, and the phrase 'REUSE AT HOME, RECYCLE. Every little helps.'*[3]

What do you do if you're the waste industry, collecting thousands of tons of mixed, contaminated or multi-layered plastics that – largely unknown to the general public – are impossible or unprofitable to recycle? You make it somebody else's problem.

Exporting waste is nothing new. Wastepaper, rags, and scrap metals have been traded for centuries. In Victorian London, for example, one 'volcano-like' dust heap that towered over King's Cross in 1826 was shipped to Russia to be used in the rebuilding of Moscow after the devastating fire of 1812.[4] Waste was and is material – like any commodity, it moves where the market is. But the global waste trade in its current form did not truly take off until the second half of the twentieth century, as consumers in the Global North began gorging on cheap goods made largely in Asia. By the 1990s, the act of making things had largely been exported – that is to say, globalised, chasing cheap labour. But, while thousands of shipping containers arrived in the West every day from Chinese ports packed to the brim with our every need (and plenty of things we didn't know we needed yet), there was little to send back. China buys far less from the West than it sells (the so-called 'trade deficit'). Rather than sail back empty, canny entrepreneurs began to fill the containers with our most abundant product: waste.

At first, these exports were predominantly scrap metal, which was desperately needed by the exploding Chinese manufacturing industry. As the journalist Adam Minter writes in his scrap travelogue *Secondhand*, although the US in particular was still producing prodigious quantities of excess metal, many American smelters were closing due to tightening environmental standards, a problem they didn't face in China.[5] Chinese waste imports continued to grow, and when the country was finally admitted to the World Trade Organisation in 2000, a metaphorical dam broke. The flow became a torrent. Between 2003 and 2011, China's scrap imports increased sevenfold.[6] The country was soon the leading destination worldwide for waste steel, copper, aluminium, paper – and eventually, plastics. By 2016, the United States alone was sending 1,500 shipping containers full of waste to China every day.[7]

It's hard to fully measure the extent to which China's economic miracle was enabled by Western waste. For China, the benefits of importing our garbage were clear: its industry needed raw materials, and there was not enough domestic supply to fill the world's ravenous appetite for Made in China goods. It was also, in some ways, more sustainable. By recycling the West's waste instead of using virgin equivalent, China prevented the emissions of untold millions of tonnes of CO_2, and the extraction of billions of tons of ore.[8] Trash flooded into regions like Guangdong, 'the Scrapyard to the World', as Minter describes it, 'a place where wealthy countries sent the stuff that they couldn't or wouldn't recycle themselves; a place where former farmers took that stuff, made it into new stuff, and resold it to the same countries that had exported it in the first place.' Countless *polan wang*, or 'junk kings',[9] got rich off this new industry, among them Zhang Yin of the paper recycling giant Nine Dragons, who became China's first female billionaire. Everything had a market: plastics and scrap electronics could be sold to Chenghai, the so called 'Toy City', where millions of plastic playthings are made, often from allegedly toxic materials;[10] or Yiwu, nicknamed 'Christmas village' for turning out more than 60 per cent of the world's festive tchotchkes. In the south, a thriving electronics recycling industry imported millions of tonnes of discarded electronics in what was known as *chengshi kuanchan*, or 'urban mining' – wreaking pollution, and, over time, helping to transform China into an electronics manufacturing powerhouse.[11]

The scale of Chinese recycling by the early 2000s is hard to even picture. In Wen'an, a formerly rural region south of Beijing, as many as 20,000 plastics processors – and an estimated 100,000 workers[12] – set up shop in just a handful of villages, sorting, shredding, and melting plastic down to feed the country's voracious manufacturing base.[13] These recyclers lacked even basic safety equipment or environmental controls. Polluted wastewater filled the nearby streams, killing off the fish; the pollution was so severe that locals started to drink only bottled water.[14] By 2011, Wen'an's Xiaobai river was so toxic that when used to irrigate farmland, it wiped out the crops.[15] Locals told of lung problems, and of men so sick that they failed military entrance exams.[16] The problem was not just associated with plastics. In Guiyu, where scrapped electronics were recycled, the ground and water was blighted with heavy metals, dioxins, and polychlorinated biphenyls.[17] According to one study, a quarter of newborns in Guiyu had elevated levels of cadmium in their bloodstreams, which can lead to cancer, osteoporosis, and kidney damage, among other harmful effects. In another, 81 per cent of children tested suffered from lead poisoning.

Crackdowns have since cleared much of the informal recycling in Wen'an and other areas of China, and concentrated it into more tightly regulated 'economic zones'. But similar conditions have been reported wherever informal recycling takes hold. I have seen it myself. In New Delhi, Akbar talked a local businessman into letting me into a plastic recycling factory, on the condition that I did not reveal their details. It was a tiny plant – one upstairs room, one downstairs – in one of the city's industrial suburbs. Concrete floor, bare bulbs, exposed brickwork, only an open window for ventilation. Upstairs, a handful of workers piled bales of plastics into a shredder, which blew fragments across the tiny room like autumn leaves. One of the workers had fashioned a thick rubble bag into a sort of headscarf. Two of the workers shovelled the flake down a chute to the ground floor, where another pair fed it into a rudimentary-looking extruder, to be made into nurdles. A half-open barrel of what I took to be masterbatch, oleaginous and green, stood in one corner. A diesel motor sputtered and churned, turning the belt. The nurdles coming out of the machine were still warm, the colour of graphite; I suspected a high haze, but there was no laboratory here to test in. It was furiously hot. The line worker's red T-shirt was dark with sweat, and the thick fug of plastic fumes in the air left me dazed. We left after a few minutes. This, Akbar said, was one of the good ones. I thought: *20,000 operations in just Wen'an alone.*

When the Chinese government passed Operation National Sword in 2018, it claimed that the ban on 'illegal foreign garbage' was 'to protect China's environmental interests and people's health'.[18] In truth, the forces that led to National Sword had been building for decades. Ultimately the reasoning was simple: China had finally grown to the point that it was no longer short on raw materials. Today it generates more than enough waste domestically to feed its recycling industry. Meanwhile, plastics pollution had become a national concern. In 2014, a documentary film by the director Wang Jiuliang, *Plastic China*, gained widespread attention in the country for exposing the reality of the plastic trade. The film, which follows a family of poor recyclers in Shandong province as they go about their daily lives, is bleak and unflinching. The family is hobbled by poverty and the father's alcoholism; in one scene, their 11-year-old daughter plays with a Barbie pulled from the rubbish. (The film was later removed from the Chinese internet by state censors.)[19] If many had known about the state of China's waste imports before, *Plastic China* helped to reveal the truth – at home and abroad.

It is hard to overstate the impact of National Sword on the global waste industry. At the time of the ban, the European Union was sending 85 per cent of its plastic exports to China.[20] Of all the plastic waste exported worldwide since 1992, China had received nearly half.[21] After National Sword, its plastics imports fell by 99.1 per cent, virtually overnight.[22] In the US, exports of plastics waste to China – all those container shipments leaving the port of Los Angeles loaded with garbage – fell by 92 per cent. Major recycling companies went bankrupt; plastics sent to landfill increased by nearly a quarter.[23] All of that waste had to go somewhere, and so almost immediately waste shipments began to flood into South East Asia: Malaysia, Indonesia, Sri Lanka, Vietnam, even uninhabited islands in the South Pacific.[24] British exports of plastics waste to Malaysia tripled; American plastics waste exports to Thailand increased by 2,000 per cent.[25] Perhaps inevitably, the new destinations included countries with among the highest 'waste mismanagement' (industry speak for dumping and burning) rates in the world.

For Yeo Bee Yin, the stories started in early 2018. 'We received a lot of complaints from a small town in the Klang valley,' Bee Yin tells me. 'Polluted water. Many people complained about the smell.' At the time, Bee Yin was the Malaysian government minister in charge of environmental issues. She has long dark hair and a politician's charming, full-toothed smile. When we speak, Malaysia and the UK are in the midst of coronavirus lockdowns, so we talk over video chat; connection, glimpsed in isolation. She's wearing a checked dress and sits in front of a window. Behind her, I can see sunlight bleaching the rooftops of Kuala Lumpur under an azure sky.

Bee Yin instructed her agency to investigate. 'We had them go have a look, and they saw that there were a lot of illegal plastic recycling factories,' Bee Yin says. 'At first we didn't know it was foreign waste.' She ordered the factories to close. But more quickly sprang up elsewhere: Jenjarom, Ipoh, Sungai Petani. 'It kept coming, no matter how many we closed,' she says. Soon there were hundreds of factories, sorting and processing plastics. Many were legal, but most were not. 'When we would go to most of these factories, they were very, very basic. They were using basic equipment. The washers that washed the contaminated plastic discharged into the river without treatment.' Plastic dumps were building up in fields and on the outskirts of villages. Locals complained of waste burning in the night.

The cheapest plastics for recyclers to buy on the open market is unsorted waste: HDPE, PP, polycarbonate, LDPE. In most cases, it is the material that waste companies consider either too contaminated or physically difficult to recycle in the Global North. But sorting is as cheap as labour, and in the Global South small reprocessing shops can scratch out a living sorting through the bales. Any plastics that can't be easily recycled are regularly sold as cheap fuel to cement factories or furnaces; in Indonesia, plastic trash was used in tofu factories, to fuel the boilers.[26] Anything remaining was (and is) dumped or burned. 'Importing the contaminated mixed plastic is lucrative if it is illegal – that means you do not need to comply with all the permits and the costs of compliance,' Bee Yin told me. 'The moment that you need to actually comply to Malaysian standards, it doesn't make sense any more.'

When illegal dumps were found, the culprits would regularly set fire to the waste instead of cleaning it up, blighting the landscape. In Malaysia, the soil at burn sites was found to contain high concentrations of toxic pollutants, including lead, cadmium, flame-retardants, and antimony.[27] This tactic – burning plastic to get rid of it, rather than pay for cleanup – is a tactic employed by waste criminals all over the world. In some cases, an 'accidental' fire might be used as the basis for insurance fraud. Others are more brazen. Cleaning up is expensive; arson is cheap.

Waste has always attracted crime. Rubbish is in many ways a perfect front: it commands substantial profits, requires only the barest untrained labour, and naturally deters all but the closest scrutiny. (No one wants to be crawling around inside garbage trucks looking for evidence of wrongdoing.)[28] In the United States, the mafia controlled the waste trade up and down the East Coast for decades: as far back as the 1950s, it was an open secret that private garbage collection was used as fronts for Italian-American crime families.[29] (There's a reason Tony Soprano worked in waste management.) Private garbage collectors run by the mob would charge extortionate rates, racking up hundreds of millions of dollars a year in profit. Any attempts to upset the garbage racket was met with warnings, beatings – and even murder.[30] As late as the 1990s, when corporate waste giants such as Waste Management Inc. started to buy up waste businesses in America, the East Coast was considered off limits. In 1992, a man attempting to set up a new waste collection business in New York opened his door one morning to find the severed head of a German shepherd on his doorstep. In its mouth was a note: 'Welcome To New York'.*

Similar tales abound. In Japan, the waste business has been exploited by the Yakuza;[31] in Honduras, landfills are allegedly used as a front by the notorious central American gang MS-13.[32] In Italy, the *capo dei capi* of waste crime, rubbish is synonymous with the mafia. In the 1980s, Italian gangs began importing waste from countries as far as Germany and Australia, and exporting it to be dumped overseas in poorer countries such as Ghana and Egypt.[33] The notorious Calabrian mafia, the 'Ndrangheta, spent many years loading ships full of illegal toxic waste, including radioactive waste from hospitals, and scuppering them off the coasts of Italy and North Africa. In Campania, the region around Naples, the Camorra mafia buried more than 10 million tonnes of toxic waste across an estimated 2,000 illegal dumpsites; the dumping is thought to have caused such a large spike in cancer cases that the region has become known as the 'Triangle of Death.[34] ('We're polluting our own house and our own land,' one mobster supposedly said to another. 'What are we going to drink?' The reply? 'You idiot. We'll drink mineral water.')

In the UK, waste crimes are so commonplace that in 2021 the Environment Agency and the police established a dedicated Joint Unit for Waste Crime (JUWC). Since 2013, the government has intercepted up to 500 illegal waste shipments annually,[35] and currently track around 1.3 million cases of illegal dumping, or 'fly-tipping', every year.[36] Among 60 organised criminal gangs being monitored for waste crimes, the JUWC

---

* It wasn't until 1995 that the mafia's trash monopoly was finally broken up, following a five-year sting by the NYPD dubbed 'Operation Wasteland'. One of the men who took credit for the sting was Rudy Giuliani, then New York mayor, now famous as . . . well, Rudy Giuliani.

has said, 70 per cent were involved in money laundering.[37] At least one criminal organisation exporting waste from the UK to Turkey reportedly used the shipments to smuggle cannabis, cocaine, methamphetamine and steroids, on the basis that customs officials wouldn't look too closely at rubbish.[38]

Today, waste management is a corporate business, controlled by a handful of large, publicly listed multinational corporations: Veolia, Biffa and SUEZ in the UK; Waste Management Inc. and Republic Services in the United States. Despite its more respectable face, Big Waste has not been without its own legal troubles. In the US, waste companies have faced charges of antitrust violations, price-fixing, and bribery dating back to the 1990s;[39] Waste Management Inc., the country's leading waste disposal provider, has settled more than 200 cases against it, most famously a $30 million lawsuit in 2005 for accounting fraud.[40]

In the UK, Biffa – whose staff so kindly showed me around the plastic factory – was in 2019 found guilty of attempting to send shipping containers of wastepaper to China that were contaminated with, among other things, soiled nappies and sanitary towels.[41] The same thing happened two years later, this time with another estimated 1,000 tonnes of 'wastepaper' (actually containing used nappies and mixed plastics) destined for ports in India and Indonesia.[42] In both cases Biffa disputed the charges, claiming the waste met import standards at their destinations. (In fact, the company's barrister argued that the waste was 'as good as it gets in the UK' – a case of damning with faint praise if ever I've heard one.) That same year, 2021, three former Biffa employees alleged that they had been trafficked to the UK from Poland and forced to work for the company by a criminal gang. According to British police, as many as 400 people had been similarly trafficked to the UK and forcibly employed in farms, slaughterhouses and recycling facilities, where they were paid as little as £0.50 per day, in what has been called the country's largest ever modern slavery case.[43] (Biffa denies any allegations of wrongdoing.) I couldn't help but think of the case when I visited another MRF, which employed so many pickers from Eastern Europe that the signage was written in Polish.

Many of the recycling factories that appeared in Malaysia and across South East Asia after National Sword were actually set up by Chinese businesses. By processing the waste in South East Asia, Chinese recyclers could export finished nurdles past National Sword, rather than waste. 'Malaysia having a big Chinese community, many can speak in Mandarin, and that makes it very easy to do for them,' Bee Yin says. (In fact, according to the academic Joshua Goldstein, China's state-backed Scrap Plastic Association had helped set up tours for businesses to scout factory locations in Malaysia, Thailand, Vietnam, and the Philippines.)[44]

Throughout 2018, the Malaysian authorities led raids on more than 130 plastic facilities, and began turning away shipments at ports, where containers full of plastics waste were piling up. Eventually, Bee Yin decided to take more permanent action. 'I thought, "If I want to solve the illegal plastics recycling factories problem, I cannot keep closing them down, because the plastic keeps coming in." So I decided that we should go to the source of the problem.' In January 2019, Malaysia banned the importation of plastics scrap, and began shipping containers of illegal plastic waste back to their countries of origin: the UK, US, Canada, Australia. That day, Bee Yin stood in high-vis in front of a shipping container full of illegal waste, and said it plainly: 'Malaysia will not be the dumping ground of the world.'[45]

Malaysia was not alone. Since National Sword, Malaysia, India, Vietnam, Thailand, Indonesia, Taiwan, and the Philippines all either announced bans or tighter controls on plastic imports.[46] Many have also begun to ship foreign waste back to where it came from. In 2019, 187 countries signed an amendment to the Basel Convention, the international treaty which governs the transboundary shipments of hazardous wastes, to include restrictions on plastic exports.[47] (The United States and China are both among the small group of countries which have not signed.) At the time of writing, at least one major shipping firm has announced that it will no longer carry plastics waste. Others will surely follow.

The international exportation of waste, meanwhile, has slowed significantly – but has not stopped.[48] As I was reporting this book, I watched as a cycle began to emerge: foreign plastics would flow into a new country, overwhelm its recycling system; the country in question would ban said imports, and then the waste would move elsewhere, to start the cycle anew: Turkey, Poland, Bulgaria, Romania, Mexico, El Salvador. No matter what new agreements passed, or how many containers were sent back, the waste kept flowing.

In the 1980s, environmental activists coined a phrase to describe the dumping of waste by rich countries on poorer countries: *toxic colonialism*. The term feels apt. Inflicting our waste on others is an act of exploitation, even domination. 'Dirt is matter out of place,' the anthropologist Mary Douglas once famously wrote – the 'right' place, of course, being something decided by those with power, and inflicted upon those without.[49] At the same time, the term *dumping*, so often used to describe the waste trade, sat uneasily with me. Waste materials are a commodity which, for better and worse, flow where markets lead.

The waste showing up across Asia and the Middle East after National Sword was being purchased by businesses – people – who presumably saw value in it. The phrase *dumping* seemed to rob those people of any agency. The truth, as it so often is, was more complicated.

I wanted to understand how waste exports worked – who was selling this waste, moving it, delivering it to those foreign ports? Who was buying it when it got there? The international waste trade is an intentionally opaque business, dominated by elusive waste brokers, who buy material from waste collectors and sell it on. But the brokers themselves are rarely named, and almost never give interviews.

The truth is that the economics of waste are set up in a way that makes the logic of exporting waste almost irresistible. In the UK, recyclers can earn money through the Packaging Recycling Notes (PRNs) scheme. To do so, they must provide evidence that packaging has been reprocessed through an accredited recycler. To claim the equivalent Packaging Export Recycling Notes (PERNs),[50] however, the traders do not have to provide evidence that the waste has actually been recycled – a full container full of plastic is considered 'recycled' almost as soon as it leaves port. The scheme is inevitably wide open to fraud, so much so that waste brokers jokingly refer to PERNs as 'Range Rover vouchers'.[51]

The actual waste itself travels the way all commodities do: it's loaded onto trucks, and from there onto the container ships that enable the global economy. Once at sea, it may stop off in any number of intermediary hubs – Antwerp, Rotterdam, Hong Kong – on the way to its final destination. How many containers of waste are there, stacked up in freeports and on quaysides around the world? Hard to say. Shipping firms rarely probe any further into the contents of containers than legally required – there are simply too many to move. The port of Los Angeles alone handles nearly 11 million containers a year. Once, I took an Uber through Long Beach, and discovered the driver was a part-time dockworker. Stuck in the city's notorious traffic, watching monolithic cranes loading 40ft containers onto a trans-Pacific megaship, I asked if he and his colleagues ever knew what was inside. I don't recall his exact response, but it might be best summarised as: *We don't know, and we don't care.*

Almost all of my efforts to speak with waste exporters fell on deaf ears. Media coverage of the waste trade was not exactly effusive in the wake of National Sword, and the promise of publicity holds little appeal for people to whom confidentiality is a primary selling point. Then there was the pandemic, which had caused a supply chain crisis in ports across the world. After a few months of trying, I wasn't getting anywhere – until I met Steve Wong.

Or rather, we started messaging. The pandemic was still raging, and Steve, the CEO of the waste multinational Fukutomi Recycling and the executive president of the China Scrap Plastics Association, lives in Los Angeles – although in normal times he spends most of his life on the road, meeting clients all over the world. A waste trader for more than thirty years, he runs an export business reportedly worth more than $100 million, and has represented the global recycling industry to the EU and the United Nations. At 65, Steve has the weathered look and restless mien of a workaholic, dark spectacles and darker hair receding gracefully. In contrast to the curt rejections or empty silences that I'd received from others, Wong was warm and obliging, happy to share the inner workings of the waste trade.

Steve was born in Guangdong, China, and raised in Hong Kong, where his father ran a small business importing plastic scrap — bag offcuts, industrial scraps from America. Steve would help out with the business after school: sorting through the waste, operating machinery, accompanying his father around the port, picking up the trade one piece of refuse at a time. 'I told myself this is not going to be something that I do for the rest of my life,' Steve says, smiling at the recollection. When he was old enough, Steve's parents sent him to boarding school in Britain. By then, plastics were in his bones.

After graduation, Steve got a job in an import/export business in the UK, and it wasn't long before he expanded into scrap. 'My first contact was a company from Leicester. He had PVC scrap, and I knew recyclers in Hong Kong,' he says. At first, he was shipping just one or two containers a month. That was the 1980s. By its peak in the early 2010s, Fukutomi was moving up to 390,000 tonnes per year, and reportedly accounted for 7 per cent of all the scrap plastics being imported into China.[52] 'Before National Sword came in, I was doing about 1,700 containers per month,' Steve says.

When Steve started out, importing waste into China was still novel. 'There was no system, no regulation on the imports,' he says. 'We had to only buy a licence.' Some importers found ways to get around the licensing system by partnering with plastics manufacturers, and piggybacking on their import quotas, 'which meant that you could bring in material without paying any import duty,' Steve explains. Abuses of the system were rife. 'There was a lot of scandal in those days. The port authority officials, their wages were only like ¥400 a month. But his authority is so big he could make you a billionaire, or a millionaire, just with his pen.'

Those kind of business practices, commonplace at the time, are no longer looked upon kindly by Chinese authorities. 'They called this smuggling,' Steve says, wearily. It remains a sensitive issue for him, legally and politically. 'Today I cannot go to China because of those days, even though I have been representing China at EU meetings.'

Today, Steve is a legitimate businessman, the CEO of a multinational whose assets were once valued at $900 million.[53] Talking to him, I'm struck by Steve's ceaseless energy. He is a serial marathon runner, father to six children, and never seems to stop working. Whatever hour I message him, he will respond, no matter the time zone. I can immediately see why he has thrived in a business built on relationship-building, trust, and hustle. (Chinese has a word, *guanxi*, which refers to one's ability to make business connections, often based on the exchanging of favours. Steve Wong has *guanxi* to spare.) Steve's openness has also fared him well, keeping him open to trade in materials that others wouldn't. Today Fukutomi trades in forty different resin types, including industrial polymers that are not recycled in household waste, but highly valued by manufacturing clients. 'Most people only recycle or recover PS, PP, HDPE. The rest cost too much money,' Steve says. 'But electronics waste might have sixty to a hundred types of polymer. The automobile industry the same, but nobody touches on it. We need a system to take care of all these unpopular materials.'

I get the sense that Steve, with enough time, could recycle almost anything, the years of knowhow stretching back to afternoons on the Hong Kong docks with his father. 'You might have DVDs at home. These are metallised. If you remove the metal, this is a really good material, polycarbonate,' he says. 'In the bathroom or in the automobile industry, you have parts which may be metallised. So if there's a chrome plating, you can sell the chrome. You can also sell silver.' Another example: airbags. 'There's a layer of silicon inside – this is a really good material. If you remove the silicon, you get pure Nylon 66. This material is worth $3,000 per tonne!'

After National Sword, Steve watched as Chinese recyclers moved to South East Asia to circumvent the ban. 'At the beginning they went there with the hope they can continue their business,' he says. 'The problems are they don't understand the language, they don't have the visa. So they can't open a bank account, they have to keep their money in other people's wallets. Some of the people lost money because they were cheated by their local partner.'

Chinese recyclers, Steve says, pay little attention to local regulations. 'They just run their place, they don't care about regulation. It's kind of a habit.' I ask about the pollution left behind – the dumps, rivers, the blighted soil. 'Recyclers are normally doing things the most financially efficient way, this doesn't mean the most environmental way.' He is, perhaps inevitably, dismissive of the pollution recycling had caused, and lays the blame for pollution at the hands of rogue actors. 'It's nothing wrong with our industry. What is wrong is the way some recyclers are doing the processing,' he says. Part of the problem is lack of investment – the price of recycled material was so low that recyclers could never invest in the appropriate machinery for their factories. The other, he says, is perception. 'The good recyclers never allowed media to visit their factories, so you cannot see a factory doing things properly.'

Steve is understandably defensive of the industry that he has dedicated his life to, and that it is his job to represent. And, even if I disagree with him on many things, there was value in what he was saying. Recyclers in the Global South might not meet with regulations, but they are serving a genuine demand. Malaysia's plastics industry alone is worth $7.2 billion, representing nearly 5 per cent of the country's GDP.[54]

Since the import bans, Steve has seen more and more waste traders going bankrupt. 'They just don't get enough material,' he says. 'Some of them focus on local recycling. Some of them moved to South East Asian countries. Some of them see that the future is gone in recycling, so they switch to other things, or they retire.' He has felt it, too. The closure of the Chinese market, and the increasingly hostile environment in South East Asia, has made the waste export business more difficult than ever in the four decades he's been working. 'We used to do 1,700 containers per month. Today, we only do 200,' he says. 'We are able to survive. But also, I work much more than a lot of other people.'

To Steve, any questions about the damages of dumping are both overblown, and also naive. 'This is a good business,' Steve says, more urgently. 'I've been to a lot of developing countries that actually love this business. They *want* to have recycling from China moved to their country. A couple of years ago, I was appointed by the government of Haiti, where I met the prime minister, and we were talking about how to build an economic zone to do recycling. I have been to Panama as well.' Moreover, he says, the Basel Convention and other bans on moving waste material ignore the simple mathematics of the situation: most countries in the Global North don't have anywhere near the recycling capacity to process their own waste, and factories in the Global South cannot source enough waste locally to feed demand. 'If all this waste only remains in developed, OECD countries, non-OECD have no chance to get their recycled content, because they don't have any infrastructure to collect the waste,' he says. 'Not only are the regulations not in place, the system is not there.' In many ways, much of the Global South is in the position now that China was in, decades ago.

Talking to Steve leaves me conflicted. He has a point about

the waste trade: we can't talk about exporting our waste without recognising the role of the people importing it, and recycling it, at the other end. What is garbage to us might be raw materials to someone else, halfway around the world. But I also can't help but feel that a lifetime in the business has blinded him to the damages being caused. If waste export was a good business, it wouldn't be in the shadows. A few weeks after my last conversation, I see reports that children as young as nine have reportedly developed health conditions while working in the plastics trade in Turkey – a country whose recycling business has grown fat on garbage imported from other countries, including Britain.[55] Maybe even my own.

Steve is right about one thing. In the UK, at least, the present reality is that there simply isn't enough capacity to recycle our own waste, and particularly the difficult-to-recycle plastics, at home. I think about Malaysian fields and Turkish ditches full of crisp packets and cat food pouches. Most of these objects do not need to exist in the first place; as Chris said, we can design away the unrecyclable materials. But if we are going to try to recycle them, then we need to have an honest conversation about where, and by whom.

Talking to Steve just reaffirms to me that we need to start telling the truth about recycling – what works, and what doesn't. That is the start of a serious conversation, one that would require us to confront the impact of our economic system, and its creation of consumption and waste in a fundamental way. But the problem is complex, and rarely black and white; any attempts to change it must involve everyone affected, wherever they may be. That will take time and political will, something which can often feel in short supply. Until then, the waste keeps piling up.

And so it is burned.

4

UP IN SMOKE

'We're burning with quite an intensity'

Six days a week, a train departs London and travels 110 miles across the English countryside to Bristol, on the west coast. The train carries no passengers. Instead, it carries seventy-eight container cars stuffed to the brim with rubbish, the collective refuse of the more than 1.6 million Londoners – from Ealing, Brent, Harrow, Hounslow, Hillingdon and Richmond – served by the West London Waste Authority. The trains are waste freight, a regular feature of the rails for more than a century; the secret garbage trucks that move out of sight, often at night, usually carrying waste from cities to rural areas, to meet their ends. Not so much waste removal as waste migration. Train watchers refer to these services as 'Binliners'.

One Friday in November, I find myself driving this particular Binliner's route from the capital to the banks of the River Severn, the wide estuary at which the west of England meets the southern edge of Wales. By pure coincidence it's Black Friday, consumerism's holy day (patron saint: Jeff Bezos), and undoubtedly the most wasteful day of the year other than Christmas. It seems fitting, then, to spend it visiting an incinerator.

The euphemistically named Severnside Energy Recovery Centre (SERC) is a modern industrial building just outside Avonmouth, with a gleaming frontage and a stepped architecture redolent of a half-spilled Russian doll. It does not look like somewhere you'd associate with burning garbage. The only sign of its true purpose is the chimney stretching skyward, so tall I can perceive it swaying slowly in the wind. It's the day after Storm Arwen has finished battering the country, and there is standing water in the gutters and ditches. Waves ripple across the Severn with the hurried purpose of pedestrians at rush hour. Upstream, I can make out the stalwart expanse of the Prince of Wales suspension bridge, and Wales beyond.

SERC, like Ellington landfill, is owned by the waste management giant SUEZ. Once again, while other waste companies have denied or ignored my entreaties to let me into an incinerator, SUEZ are more than happy to help out – and so here I am. Technically, SERC is not an incinerator but an energy-from-waste plant (or waste-to-energy, depending on which company is doing the naming). The distinction is an important one: while incinerators of old were purely a means to dispose of waste, energy-from-waste plants are modern utilities. This massive plant burns 430,000 tonnes of waste a year, and in so doing produces up to 40 megawatts of energy for the UK's national grid. 'Enough to charge every phone in the UK at the same time, or so I'm told,' Andrzej Posmyk, one of the plant's senior engineers, tells me. Andrzej is young, fresh-faced, with a sweep of black hair like he should in a boyband. He's been tasked with giving me the tour. I'm also accompanied by the plant's delightfully named operations manager, Thomas Merry, but Thomas is suffering from vocal palsy (long story) and his voice is hoarse. 'It doesn't help that I was up for two hours in the middle of the night last night,' Thomas, croaks, gamely. He has a nine-month-old baby son who isn't sleeping. I can relate, my youngest daughter being even younger and similarly insomnolent.

After the usual safety ritual (PPE, helmet, earplugs) we head down in a lift and out into the plant. Most of the waste that arrives at SERC – around 330,000 tonnes a year – arrives on the Binliners from London. The railhead is right outside the factory, overshadowed by container cranes. 'It's mainly black bag waste,' Andrzej says, meaning that SERC burns mainly mixed rubbish, that which does not pass through a materials recovery facility. 'At that point, we're really relying on the public to know recycling.' The remainder of the waste burned inside SERC is commercial, although it can be 'supplemented' from local waste collections, when fuel supplies are low. 'There's always demand,' Andrzej says. 'The amount of waste everywhere is increasing all the time.'

It's likely that humans have burned waste for millennia. Fuel is hard to come by; trash burns. The first municipal garbage incinerator, known as a 'destructor', was patented by Alfred Fryer, a sugar refiner in Nottingham, England, in 1874.[1] At the time, burning waste was seen as both cheaper and more sanitary than the dust heaps and dung heaps that plagued cities. (This was, remember, the height of miasma theory.) In 1876, Fryer's design was adopted by the city of Manchester, and other industrial towns and cities quickly followed. It wasn't long before intrepid engineers realised that incinerators could be harnessed for other purposes: in Preston, a tram system ran on steam from the local destructor; in Woolwich, waste powered the town's street lights.[2] Destructors – which solved the waste problem, while providing an economical means to generate electricity – were quickly adopted throughout the industrialised world, particularly in the United States, whose engineers viewed the English invention with envy. By 1914, according to the garbage historian Martin Melosi, approximately 300 incinerators (they liked the idea, but apparently not the name) were operational across the US and Canada.[3]

Not everyone was keen on the idea of burning their rubbish. In London, some critics viewed incinerators as a 'wicked and wilful waste of money', in that they might destroy 'valuable refuse'.[4] To compound matters, the first generation of incinerators were plagued with technical issues. Wet waste – food waste included – did not burn well, and so the plants had to be topped up with coal, increasing operating costs. They also left behind significant amounts of ash, known as clinker (not to be mistaken for the cement ingredient), which had to be disposed of separately. As a result, many towns simply reverted to landfill. By 1920, dozens of incinerators across the US had closed down.

Still, the idea of burning trash is tenacious. After the Second World War, home incinerators, known as 'burners', became fashionable household appliances, particularly in rural America, where waste collection was unfeasible.[5] By 1957, the popularity of these back garden incinerators was thought to be contributing so much towards the smog problem in Los Angeles that the city banned their use completely.[6] It wasn't until the 1970s, with the environmental movement growing and concerns about energy independence caused by the Oil Crisis, that a second wave of large-scale incinerators arrived, now rebranded as energy-from-waste (EfW) plants.

Today, in many countries, incineration has replaced landfill as the default destination for household waste. In the UK, the percentage of waste that ends up burned has grown from 9 per cent in 2001 to 48 per cent in 2021 – a 435 per cent increase.[7] In the European Union, which burns just over a quarter of its trash, incinerators power 18 million homes.[8] Sweden burns roughly 50 per cent of its waste, Japan 78 per cent.[9] Denmark, which burns four-fifths of its household waste, has built so many EfW plants that that the country now has to import around 1 million tonnes of waste per year to keep them running effectively.[10] China, which has built more than 300 EfW plants since the 1990s, burns 580,000 tonnes of rubbish every single day.[11]

At Severnside, cranes unload the rail containers onto trucks, which lug the waste inside and onto the cavernous tipping floor. I watch from an overhead gantry as the trucks unceremoniously dump their cargo into 'the pit' – a cavernous trench more than 21,000m³ in capacity, or larger than eight Olympic swimming pools stacked together. The pit already contains a truly immense amount of rubbish: burst garbage sacks; ribbons of plastic; soiled clothing; old furniture; shattered wood. 'It can store 12,000 tonnes,' Andrzej says. I ask how far the down the waste goes. 'I haven't seen down to the bottom before, but it is a deep, deep, *deep* pit.' It's grotesque to witness, like one of Ghazipur's trash mountains inverted in the earth. Looking at it, I feel the same sensations that I had in India: repelled and compelled. Do we hide our waste because we are disgusted, or ashamed?

We cross a walkway into SERC's main control room, where around half a dozen staff are quietly monitoring the plant's systems on banks of computer monitors. The room is painted a sunny shade of yellow, and surprisingly pleasant: there's coffee on in a kitchen in the corner, David Gray playing on the radio. This, I imagine, must be to counteract the unpleasantness outside. On one side of the room, windows look out into the pit through a wire safety fence, the glass caked with grime. Two gigantic grabbing claws are suspended above the pit, each large enough to lift a minivan. An older technician named Wayne is operating the cranes from a chair armed with twin joysticks and a series of fun-looking buttons.

It looks like a fairground claw machine, I say.

'Pretty much the same, yeah,' Wayne says, unfazed. 'Do you want a go?' Of course I do – but to my disappointment, Thomas intervenes. Apparently letting journalists operate heavy machinery with no training is against protocol, and so I have to settle for a demonstration. 'It's pretty straightforward. You've got forward, backwards, right, left,' Wayne says, pulling at the controls. The cranes move slowly, Damoclesian, their noise a deep rumble.

The cranes are designed to lift waste from the pit and drop it down two large hoppers, where ram feeders slowly push the waste into the furnace. At SERC, unlike the previous generations of energy-from-waste plants, much of the combustion and processing is automated. 'These cranes will drive themselves,' Andrzej says. 'Every now and then if we need to do something in the pit, one of the guys jumps in the chair to move some material manually.' I squint through the glass. From this aerial view, individual objects are almost indistinguishable among the refuse. I can make out some polystyrene crates in sharp contrast with the black-bagged mass. Otherwise, it has coalesced into something homogenous, other. The sensation of looking down at it is not unlike that of looking into a toilet.

Some wastes are burned for a reason. Medical waste, for example, invariably contains infectious pathogens, used needles, and countless pharmaceuticals, which improperly disposed of can present a danger to people and to the environment. Opiates, hormone treatments, cancer drugs – these are not things that you want leaching from a garbage dump. Antibiotics left inside landfills have been shown to promote the growth of antibiotic-resistant bacteria, which if released risks polluting soils and waterways.[12] Then there's anatomical waste, a category that includes blood and body parts.

In the UK, medical waste has historically been incinerated, although that is not without problems. In 2014, an investigation by the TV series *Dispatches* found that more than 15,000 unborn foetuses in British hospitals had been treated as 'clinical waste' and burned, sometimes in energy-from-waste plants; the hospital in Cambridge where both my children were born, and in which my wife Hannah and I once suffered a miscarriage, treated 797 foetuses in this way. The parents were told their babies would be 'cremated'.[13] (After the story broke, the practice was banned.)

The tight regulations in the UK over how and where medical waste can be burned can create bottlenecks. In 2018, an NHS contractor called Healthcare Environment Services was found to have stockpiled hundreds of tonnes of anatomical waste in warehouses in England and Scotland. The company blamed a lack of incinerators with sufficient pollution controls. (A criminal investigation is under way, and the company has since gone bankrupt.)[14] Several countries have now decided that even medical waste should not be burned. The United States, Netherlands and Germany have all begun to phase out medical incineration in favour of using autoclaves (in which medical waste is sterilised with high-pressure steam) or microwave treatment, and sending more medical waste such as plastics for recycling.

* * *

The crucial measure in energy-from-waste is called the Calorific Value, 'CV', which describes how much energy the waste contains at any given moment. 'There is variability in how well the waste burns, depending on whether there's more plastic in there, more paper, someone decided to throw a log of wood in, metals – those are sort of things that we have to try to account for,' Andrzej says. Plastics and wood burn well, while inert materials like ceramics slow down combustion. Leachate also accumulates at the bottom of the pit, making the deepest waste heavy and wet. To counteract this, the pit must be regularly mixed; the action in practice looks a little like tossing a salad. 'In a perfect world, we'd be mixing and dragging material off the bottom of the bed, but that's not possible,' Andrzej explains – there's simply too much waste, and more material arriving all the time.

The biggest challenges for the crane operators is 'off-spec' waste. 'You might get something like a huge tree stump, or gas canisters that can pop off and cause problems,' Andrzej says. The issue is worst during the holidays and sales. Black Friday – today – will inevitably lead to a huge short-term spike in waste, particularly packaging. 'The biggest influx of waste is the week after Christmas. The amount of waste is huge,' Andrzej says.

What is the strangest waste that comes through, I ask?

The room fills with chatter, workers swapping tales. 'A giant concrete block, half the size of you,' Andrzej says.

'The front of a car,' says Thomas.

'We had an anchor and chain come all the way through at one point.'

'A dolphin carcass,' Thomas says. 'Washed up on the beach near here. That went through.'

In the control room, monitors show a live feed from a thermal camera inside the furnace, an indiscernible blur of white heat. An indicator shows the current temperature inside — 1,027°C, although it can reach 1,300°C. 'We run it at such a temperature to burn off any harmful chemicals,' Andrzej says. Burning waste emits a spectrum of pollutants, ranging from heavy metals like mercury, cadmium, and lead, to hydrochloric acid, a major constituent of acid rain. Burning waste is notorious for emitting dioxins and furans, two highly toxic chemicals known to disrupt hormones and cause a range of cancers. Dioxins and furans are among a class of toxicants known as persistent organic pollutants (POPs), which can take decades to break down in the environment, and so accumulate; ingested or inhaled, they can interfere with the immune system, the liver and the brain. The two chemicals have long been associated with the incineration industry. The United Nations Environmental Program once concluded that prior to 2000, incinerators were responsible for 69 per cent of worldwide dioxin emissions.[15] However, following a slew of air quality legislation in the early 1990s, modern EfW plants must follow strict limits on the levels of dioxins and other pollutants that they can emit. The result has been a fall in dioxin emissions of 99 per cent. Today, the UK's incinerators supposedly emit less dioxin every year than Bonfire Night.[16]

Even so, incinerators' emissions are substantial: in the UK, energy-from-waste plants have been found to emit more CO_2-equivalent per kilowatt hour generated than coal power stations,[17] while regulators in New York found that even incinerators that comply with air quality legislation can release up to twice as much lead, four times as much cadmium, and up to fourteen times more mercury as coal.[18] It's not just CO_2. Italian researchers found that living near waste incinerators was associated with an increased risk of miscarriage;[19] in China, people living close to waste incinerators faced higher exposure to heavy metals and airborne particulates, increasing their chances of long-term health defects such as cancer.[20] In London, pollution emitted by the city's five EfW plants may contribute towards fifteen deaths per year.[21] Incinerator emissions are also likely to be higher in older plants, and in countries with more lax regulations; walking around SERC, I couldn't help but think of the abandoned incinerator at Ghazipur. (Several other studies, by contrast, have found that incinerators pose little to no direct health risk. A study in the UK of more than 1 million births found no link between living near incinerators and infant mortality; although a follow-up did find a very small increase in some birth defects, the researchers could not prove causality.)[22]

In the control rooms at SERC, the banks of displays show detailed readouts of the emissions from the stack at any given moment. The furnace emits 200,000 cubic metres of flue gas every hour. First the flue gases are first treated with ammonia to remove any nitric oxides (NOx), and then pass into the boiler, where they heat water into steam, used to power the turbine. The remaining gases then pass through the air pollution control system, a massive and intricate series of pipes, tanks and injection chambers designed to remove environmental pollutants. 'We put in lime, which scrubs out the acid gases – hydrogen chloride, sulphur dioxide, hydrogen fluoride,' Andrzej says. 'Then we also inject activated carbon, which removes any of the trace heavy metal content as well.' The numbers flicker up and down on the screen in front of me: nitric oxide, carbon monoxide, sulphur dioxide, organic volatiles. 'We have a set of emissions compliance limits that we have to abide by at all times,' Andrzej says. 'If we do not comply, we shut the plant down.'

* * *

Andrzej and Thomas lead me out of the control room and into the plant. The incinerator itself is Swiss, based on technology from the incineration giant Hitachi Zosen Inova. We wander through the pipes and valves, all clean and shining and meticulously arranged, until we reach the furnace itself. There's a small viewing port, shielded by thick glass. I gaze into the fire. The first thing I say at this moment, according to my recording of the incident, is: 'Oh my god.'

Then: 'Wow.'

Fire fills my field of view. The grate is covered in a heaving grey-black mass of waste, all aflame. I watch the garbage conflagrate and burn, rubbish bags melting and twisting apart, wisps of polythene ash caught in the updraft. I can feel the heat on my face through the glass. The fire appears to be *writhing*. This, Andrzej explains, is due to the ram feeders constantly feeding in new material onto the grate. The disintegrating waste drops down a series of ledges, like a penny-pusher at a fairground, until all that's left is ash and gases.

'We're burning with quite an intensity,' Andrzej says.

There's a story, almost certainly apocryphal, that the citizens of ancient Jerusalem dumped their garbage outside the city in the valley of Gehenna. The waste burned constantly, and over time the word Gehenna became synonymous with hell. Staring into the flames, I understand why. An inferno renders categories meaningless. To a fire everything is fuel; language burns. 'The guys use a special language to do fire checks, so they can refer to specific sections,' Andrzej says. 'I still haven't got the lingo. I think it's Welsh.'

Firefighters describe a phenomenon known as *flashover*, the moment when the temperature and air conditions within a room reach the point at which all combustible material will spontaneously ignite. A space ceases to contain a fire, and instead *becomes* a fire. I think that image comes to mind because even here, the waste inside the furnace is not some abstract fuel but so evidently *stuff*, the things that we have made, and used, even treasured. Some cultures say that fire is cleansing. Knowing what we know now about greenhouse gases, about the metals and persistent pollutants that are left behind in the ashes, I'm not so sure. This is fire as metamorphosis, a useful sleight of hand, exchanging one pollution for another. *Poof — watch your waste disappear.*

Some waste remains. The dust and cinders left behind in the furnace, called bottom ash, eventually tumbles through the grate and into a conveyor full of water, where it is quenched and cooled. 'We ensure that it's fully burned off, and then we send it off for processing,' Andrzej says. Bottom ash makes up the majority of waste left behind after incineration. It mostly consists of material that does not combust, such as glass, and various minerals, but can also contain heavy metals, brominated flame retardants, and per- and polyfluoroalkyl substances (PFASs) — albeit not in high enough concentrations for most countries to consider bottom ash hazardous.[23] The metals in bottom ash can be extracted and recycled. By one estimate, Europe's waste incinerators leave behind around 1.5 million tonnes of iron every year, enough to construct twenty-eight new cruise ships.[24] I ask Andrzej where the metals come from. 'Gold, jewellery, currency, but mainly electronics,' he says.

SERC produces 102,000 tonnes of bottom ash every year. It is sold to a company which extracts any recyclable metals and converts the remaining material into aggregate for the construction industry. Later, we pass a huge pile of ash outside the plant, awaiting treatment. 'It's used for things like road construction, building, creating breeze blocks,' Andrzej explains.

The ash that is carried aloft up the stack is technically

referred to as Air Pollution Control residue, and more commonly known as fly ash. Uncaptured, it can also contain heavy metals, as well as dioxins, POPs, bromines and polycyclic aromatic hydrocarbons (PAHs). (It's bad for you.) At SERC, most of these pollutants are removed in the flue gas treatment process. Even so, intercepted fly ash must be disposed of safely. Exactly how remains controversial. In 1986, a cargo ship, the *Khian Sea*, set sail from the United States for the Bahamas carrying 15,000 tonnes of bottom and fly ash from incinerators in Philadelphia. After environmental campaigners informed the destination of its toxic contents, the Bahamian government refused the boat entry. For more than a year, the newly nicknamed 'Ash Boat' wandered the Caribbean looking for a port that would accept it – Puerto Rico, the Dominican Republic, Jamaica, Honduras, Panama, Bermuda, even the Cayman Islands. All refused. In January 1988, the ship managed to dump 4,000 tonnes of ash onto a beach in Haiti, after tricking a local firm into believing that it was 'topsoil fertilizer'.[25] Greenpeace tipped off the Haitian government, and the ship fled. Eventually the *Khian Sea* crossed the Atlantic, stopping off in the likes of Morocco, Senegal, Yugoslavia, and Singapore. It tried changing its name to the *Felicia*, then the *Pelicano*, but to no avail.[26] By the time the ship finally did dock in Sri Lanka, in November 1988, the incinerator ash had mysteriously disappeared. The ship's captain later admitted he had dumped the cargo into the ocean.[27]

Some countries still ship their ash elsewhere. Since the 1990s, Norway, Sweden and Denmark have shipped more than 500,000 tonnes of fly ash every year from their EfW plants to Langøya, a tiny island outside Oslo, where it is mixed with concrete and tipped into a disused quarry.[28] The pit has nearly reached capacity, triggering a political crisis over how to dispose of the ash in future. (One option: shipping it to a former salt mine in Germany.)[29] Elsewhere, incinerator ashes are often landfilled, either in dedicated sites or mixed in with municipal waste. In the US – where bottom ash and fly ash are typically combined into one – ash is actually used by landfills as top cover.[30] In the UK, which produces around 450,000 tonnes every year,[31] fly ash is largely sent to hazardous waste facilities, underground storage, or is recycled. 'We send it to a chemical recycling company called Castle Environmental, who treat it and then use it for construction material,' Andrzej says, casually. In this way we have progressed from raising cities upon waste to building them with waste, the legacy of consumption forever suspended in the walls around us.

Burning waste is divisive. Whenever a new incinerator is built, its owners inevitably face resistance from campaign groups who argue that energy-from-waste plants are not only polluting, but a waste of taxpayers' money. They have a point. As a means of generating electricity, waste is not particularly efficient. The US's seventy-five incinerators burn around 34 million tons of waste a year, but currently produce only 0.2 per cent of the nation's energy.[32] In the UK, the figure is a little higher: in 2020, EfW plants burned nearly half of all household waste – over 14 million tonnes – but produced just 2.5 per cent of the country's energy.[33] Despite this, many countries still classify burning waste as 'renewable',* a fact the energy-from-waste industry has used to cash in hundreds of millions in government subsidies. In the UK, incinerators are exempt from the national carbon trading scheme, which charges polluters for

---

* A reminder that 'renewable' energy does not always mean 'clean' energy, despite the fact that the terms are sometimes used interchangeably.

burning fossil fuels (despite burning large amounts of plastic).[34] In the US, twenty-three states – including New Jersey, Massachusetts, Maryland, and Oregon – treat incinerators as renewable,[35] and therefore eligible for various subsidies and tax breaks. So too, until recently, did the European Union, all of which has led environmental groups to declare that the governments are 'burning money'.[36]

The problems are compounded by the way that energy-from-waste plants are built and paid for. In the UK, dozens of EfW plants have been built using so-called 'private finance initiatives', in which waste companies loan the government the money to build incinerators in return for lucrative long-term contracts paid back over decades.[37] SERC itself was commissioned under a £1.4 billion, twenty-five-year contract between SUEZ and the West London Waste Authority. The contracts often require the local council to guarantee a certain supply of waste (fuel) to keep the incinerators burning – and profitable. Some argue that these contracts prevent them from increasing recycling rates, as they are contractually required to continually feed the incinerators. In 2014, a Parliamentary report published by the British government criticised the way that incinerators had been funded, claiming the contracts were exploitative and had wasted hundreds of millions of pounds of taxpayer money.[38] (In one case, a council in Norfolk had to pay £33.7 million to get out of an EfW contract for a plant that was never even built.)

The reality is that incineration directly competes with recycling. Research has shown that countries and cities that burn the most waste also recycle less – Japan, for example, which incinerates 78 per cent of its household waste, has a recycling rate of only 20 per cent.[39] London burns the largest share of its household waste in the UK, and has the lowest recycling rate; London's Westminster council sends over three quarters of its waste for incineration, and recycles just 23 per cent. In some cases, even waste that families have carefully sorted into their recycling bins is burned. The reason for this is simple: incinerators need fuel, and many recyclables are high in calorific value. Plastics, being essentially fossil fuels, burn particularly well. (In fact, the plastics industry are one group that seem to particularly love incinerators. In the US, companies such as Dow Chemical – supported by our old friends Keep America Beautiful – have funded efforts to funnel plastic waste into incinerators so that they can be burned, rather than recycled.)[40]

'People put it in their black bins,' Andrzej says, 'and you know some things, like film plastics, are not recyclable with current technologies. So that contributes a lot of the calorific value of the waste.'

You need those materials to burn, I say.

'Absolutely, we need some of it,' Andrzej says. 'But these things' – the furnaces – 'are quite robust. They can be designed for lower calorific values as well, and biogenic content can make up for that. You know, you can burn more paper.'

Some incinerators have started to introduce new technology to increase their efficiency, with the intent to reduce their carbon emissions. Combined heat and power (CHP) plants use leftover heat from the furnace to power nearby homes and businesses. The $525 million Amager Bakke incinerator in Copenhagen, designed by the architect Bjarke Ingels, features a slanting roof covered with a synthetic ski-slope and climbing wall, which have turned it into a major tourist attraction. Norway's capital Oslo is, at the time of writing, constructing the first energy-from-waste plant that includes a carbon capture and storage facility, which will capture up to 400,000 tonnes of CO_2 per year.[41] 'It's something that the whole industry is looking at,' says Andrzej.

But doubts over its sustainability are causing some countries to re-evaluate incineration, even as others – like the UK, US, and China – double down. The EU recently disqualified energy-from-waste plants from certain sustainable funding and grant initiatives, arguing that they are incompatible with its Net Zero and waste reduction targets. Denmark, one of the world's most prolific burners of garbage, has decided to cut its incineration capacity by 30 per cent, declaring the practice incompatible with its recycling plans. Not for the first time in the last century, the practice of burning our waste is coming under question, though this time less for reasons of pollution than something more fundamental: is it the right thing to do?

Andrzej struggles with some of the criticism. 'It's really frustrating,' he says. 'The reason I went into engineering is I want to do environmental stuff. The waste industry is seen as not caring about the environment, but it's the fundamental aspect of it. I think people forget. They don't think about: "Where does the waste go?"' The reality, he's saying, is that not all waste is recyclable, or reusable. The question is what you do with the rest of it. 'At this period in time, there would be no other place than landfill for this,' he says, gesturing towards the pit. So we're left with a simple choice: do we want to bury our waste, or burn it?

We climb down to the ground level of the plant, and head out towards the railhead. On the plant floor, Andrzej shows me the concrete block that they pulled from the pit, big enough to total a minivan. As we head towards the exit, the tangled nest of pipework overhead, I spot a bird. It's elegant, black and white with a spit of tail feathers – a pied wagtail perhaps, or a long-tailed tit. I'm a poor birder, and didn't catch a picture.

Anyway, we watch it for a moment, hopping to and fro, this little bird lost inside this industrial behemoth, the furnace roaring overhead.

As cosmic metaphors go, I think, *it's not exactly subtle.*

5

USED

'The lowest shitty ones go to everyone else'

Royston, the town where I live, is a small place with a rich history. Underneath the town square is a medieval cave thought to have been frequented by the Knights Templar. The local heath, a site of special scientific interest rich with rare flowers and ground-nesting birds, is home to several Bronze Age burial mounds. King James had a palace here, for when he came to hunt grouse. But, like many small British towns outside of London, it's fair to say it has seen better days. The once bustling high street has been gutted by the rise of super-markets and online shopping. Empty units linger, boarded up. Closing-down sales are becoming more common. Only a few holdouts remain: a bakery, two opticians, a couple of pharmacies, a knitting shop. The haunts of the elderly. There are, however, no fewer than five charity shops, meaning that on an ordinary weekday your conscientious shopper can take their pick between supporting children (Barnardo's), animals (Woodgreen), elderly care (Age UK, The Garden Hospice) or the emergency services (Air Ambulance).

In this, Royston is not alone. There are now 11,209 charity shops in the UK, according to the Charity Retail Association.[1]

That figure has risen steadily since the 2008 financial crisis, even as other shops have shut by the thousand.[2] Not that long ago, charity shopping (or 'thrifting') came with all sorts of negative connotations tied up with poverty and class. Today, buying things second hand is if anything *more* fashionable than buying things new, the subject of glowing magazine features and championed by social media influencers. Some of the biggest charity chains in the UK, such as Oxfam and Shelter, have even set up destination stores for their highest-end donations, somewhere for the eco-bougie to buy Prada or Margiela guilt-free. Early in my career at Condé Nast, it was a closely guarded secret which of the local charity shops would receive the fashion magazines' unused press freebies – a secret that penny-pinched interns, often working for little more than minimum wage, would share with only the closest of friends, so that they might get in on the drop early.

We do not instinctively think of charity shops as part of the waste industry. To even say so feels somehow sacrilegious; they are supposed to be altruistic, feel-good places, somewhere to show our generosity by giving away that which others might put to good use. But that is naive. The awkward reality is that charity shops are not just waste infrastructure, but an increasingly essential part of it. In fact, in small towns like mine, high streets have stopped being primarily a place to buy goods, but more often a place to dispose of them. Consider: only between 10 and 30 per cent of second-hand donations to charity shops are actually resold in store.[3] The rest disappears into a machine you don't see: a vast sorting apparatus in which donated goods are graded and then resold on to commercial partners, often for export. This work is commonly done by the elderly, and for free – of the 213,600 people working in UK charity shops in 2022, 186,800 were volunteers.[4] We hear all the time about the benefits of the second-hand trade: in the UK alone, donations make more than £330 million per year for their parent charities, saving 339,000 tonnes of clothing from landfills and incineration, and 6.9 million tonnes of CO_2.[5] But we rarely see the other side, the networks of people involved in processing, reselling, and eventually reusing the things we give away – networks which encircle the globe like a ball of yarn, conveying our unwanted tat to people who need it, in Afghanistan or India or Togo. Like anything we put in the bin, they are sent 'away'. In this case not thrown, but given.

I first met Oleksii Kotyk at a recycling conference. There, as suited executives from the big waste firms gave talks about the circular economy while their stalls handed out single-use plastic tat like candy, and gruff waste dealers admired massive German-made shredders and track shovels like toddlers in a toy shop, I spotted a small stall, barely big enough to contain the bright pink donations box at its centre. There I found Oleksii, the founder and director of Pink Elephant Recycling, a textile recycler based in Leicester. A serious man with a square face, framed with half-rimmed glasses, he is a passionate believer in the circular economy, and in the value of reuse. He is also, unlike many, totally open about what actually happens to our donations. 'What we're doing is the real circular economy,' he told me. Did I want to come to their plant and see it for myself? Did I ever.

Pink Elephant Recycling is a deceptive place: a small red-brick two-storey building on a somewhat rundown-looking street opposite a B&Q. There's the garage, painted fuchsia, and two bright pink textile collection bins in the car park, but otherwise you would never guess the scale of the operation inside.

Oleksii meets me at the door on a chill winter's day, wearing a black hoodie and a black beanie pulled down to his eyebrows. After he loans me a high-vis vest, we head into the warehouse, a brightly lit and somewhat ramshackle space. There are clothes and commotion everywhere, like a discount store on sale day: T-shirts piled high in bales, skirts spilling from blue plastic bins, animal print onesies trying to escape their cages.

Pink Elephant processes around 2,000 tonnes of clothing every year, which makes it a relatively small player in the market. (Textile Recycling International, the largest processor in the UK, handles 130,000 tonnes.)[6] 'It's a straightforward business. You acquire material, you process that material, and you sell the material,' Oleksii says. He often calls the clothing that – raw material – as if he's moving pallets of pork or iron ore. The clothes themselves are donated via pink collection boxes around the UK. Oleksii used to take clothes from the charity sector, but stopped a few years back due to the poor profit margins. 'If you rely on the charity sector alone, you're open to being exploited,' he says. The charity shops tend to keep the best material to sell themselves, he explains, while 'the lowest shitty ones go to everybody else'. Similarly, door to door collections – in which charities leave plastic bags on doorsteps, and offer to collect them full – didn't yield enough of a return, as too few people follow through. 'So now we do textile banks, and only textile banks.'

The collections arrive in a loading bay at the back of the building, where they're unpacked and pre-sorted by garment type: men's trousers, women's blouses, kids' shoes. We pass by neatly organised cages full of handbags, men's shirts, a pool of shoes that looks deep enough to swim in. Oleksii never knows what will arrive mixed in with the donations: food, grease, actual garbage. Sometimes the bins themselves leak and the clothes moulder, at which point Pink Elephant can't rescue them. A big part of his goal is to reduce the amount of leavings sent to the incinerator — any item he can rescue from being thrown away can not only be sold, but saves him from paying disposal fees, doubling his margin. 'Our recycling rate is 90 per cent,' Oleksii says, proudly. To that end, he has recently installed a room full of professional washing machines and tumble dryers, to rescue what he can. 'What was rubbish becomes the recyclable product.'

After pre-sort, the cages are brought into the factory proper, to the grading tables, where I count at least a dozen workers ploughing through piles of donations. 'This is the trousers table, so that's only going to get trousers. Asian dress. Africa table, that's summer stuff. Heavy table — jumpers, jackets, sweatshirts,' Oleksii says, as we walk the line. At the tables, the items are categorised not only by item type, but by quality. Each category has its own sorting bin. 'Men's shorts A [grade], men's shirts A, ladies' cotton blouse A,' Oleksii says, rifling through the bins to show me, stopping politely short of women's delicates.

Most of the workers here are Eastern European. 'They're Latvians, Bulgarians, Polish,' Oleksii says. Oleksii himself is from eastern Ukraine, still audible in his thick accent. He breaks off our conversation intermittently to bark at his employees or answer his phone in Russian; going by body language and the occasional mentions of currency, I infer that deals are being made. Oleksii originally got into this business exporting truckloads of textile waste to and from Eastern Europe. 'I never even saw the material,' he says. 'Like a commodity. Buy, sell.' It was only later that he realised that what was being donated was only a thin slice of the clothes we buy as a society every year. The rest were still being thrown away.

Some might see that as a tragedy. Oleksii saw it as business opportunity.

We stop at a table to watch Anya, a blue-haired woman with purple nail polish and thickly painted eyebrows, sorting through a pile of womenswear. She picks out a tank top, pulls the fabric taught between her fingers, glances at the label, her eyes scanning for major stains or tears. 'First, we check the colour, the condition. Is it long sleeve, short sleeve? They go in different bins. This one is not good condition, so it's got B grade,' Anya says, tossing it in the bin. She pulls another: a turquoise T-shirt from Nutmeg, the in-house brand sold by the supermarket chain Morrison's. Anya rubs the material as if feeling for a stuck page. In the bright fluorescent lighting, the thin fabric looks almost sheer. 'This has lost its shape,' she says. 'B' grade. The assessment took perhaps two seconds.

'They need to know eighty-five different categories on the table,' Oleksii says. 'Taking into context they're sorting about a tonne a day each, a lot of decision making has to be done, based on the categories and the condition of the garment.'

I notice Oleksii has taped pictures of various brand logos to the tables as a visual aid. 'I have the list that a charity – I won't tell you which one – gives to their employees, of over 600 brands,' he says, conspiratorially. He pulls up the document on his phone: an Asos-to-Zegna of every clothing label I've heard of and more, alphabetised and ranked by value. 'They put them into bronze, silver, gold,' he says. That was too much for even seasoned workers to memorise, so Oleksii has slimmed his own ranking down to around 200 or so. (Broadly speaking, supermarket brands are the lowest quality, designer brands the highest.) Most of the time, it isn't needed. The vast majority of clothing that comes through Pink Elephant come from just a handful of major supermarket and high-street brands. 'Cheap fast fashion,' he says. 'Primark, Tesco, Sainsbury's.' Oleksii rarely even pays attention to 'designer' labels, due to the impossibility of authentication. 'Ninety per cent of the branded stuff is fake,' he says. 'We don't have time to sit down and count the stitches on a Chanel bag.' It all gets thrown in together.

Oleksii has been in this business for fifteen years, and has seen first-hand the decline in the quality of clothing being bought and sold in the UK. 'Before 2010, the supermarkets were not in this industry. The only [cheap] stuff was Peacock, Primark, a few other brands. But when the supermarkets moved into this fast fashion, they changed the game dramatically,' he says. The numbers back him up: according to the Ellen MacArthur Foundation, the amount of clothing bought in the UK per person has doubled since 2000, but the number of times each item is worn has fallen by 36 per cent.[7] At the same time, Oleksii says, there has been a tangible drop in the average quality of the clothing we buy. 'You're talking about density, weight of fabric,' he says. 'Some T-shirts, you can literally see through them. You take an old GAP T-shirt or a polo shirt from Ralph Lauren, even after two years there's plenty of life in it. Today, brands are using the same amount of material to make two or three T-shirts, so they are a much lower quality.' Some T-shirts from fast-fashion brands, he says, won't last three washes.

I watch the sorters work. There's something mesmeric, I find, in watching people who have mastered a task to a point beyond muscle memory, where it becomes instinctive, fluid, like watching a concert musician or a line chef in full flow. Dance music plays on the radio, keeping time. Another woman picks out a baby's sleep suit with a thick dark stain down it. It looks unwearable – but no, it goes straight in the bin, as a B grade.

'That was quite soiled,' I say.

'Baby is baby,' Oleksii says, unfazed. I'm sceptical; I wouldn't dress my daughters in a stained outfit, and it seems wrong to expect others to. 'Some marks and so on we can let through. You have to mix – you have to have enough decent stuff with enough life left in them.'

Oleksii is no longer shocked by what people donate, or throw away, but the profligacy is quite stunning. In the centre of the warehouse is a bathtub-sized cage full of children's 'B' grade jackets. 'Just look at this,' Oleksii says, showing me a blue snow jacket that looks almost new but for a broken zip. 'This is B grade, each item is 15p.' Oleksii takes a zipper off another coat, fixes the zip, and throws it into another bin. 'Now this one can go in A grade, which is five times more expensive.'

Oleksii recently bought out an old dry cleaning business, and with it a bunch of sewing machines and other machinery. 'We are planning to set up a repair space,' he says. Lately, textile recyclers have started setting up stores on eBay and resale apps like Vinted and Depop to take advantage of the soaring popularity in second-hand clothes. According to the US-based resale site thredUP, used clothing sales are expected to double by 2026, most of that from peer-to-peer selling apps. As a result, second-hand traders from the UK to Malaysia are finding themselves quickly becoming retailers, too.[8] 'I'm slowly building up my online presence,' he says.

Right now, almost all of the material at Pink Elephant is sold for export. Some of Oleksii's winter wear will be shipped to Eastern Europe; the lower grades will be shipped to Pakistan, along with the unwearable material, which will likely end up shredded and recycled into rags or shoddy.[9] But by far his biggest market is West Africa. 'Ghana and Togo,' Oleksii says. In the UK, 70 per cent of used clothing is exported. In 2018, that amounted to more than 395,000 tonnes, cumulatively worth £451 million;[10] only the United States exports more. Globally, the trade is worth £3.6 billion.[11]

In the centre of the room, three men are pressing the clothes into bales, which are weighed and wrapped in plastic. Textile bales come in various sizes and weights. 'An Africa West standard bale is 55kg,' he says. 'Some of them are done by pieces: men's trousers, 200 pieces. Lady's mixed pants, 200 pieces.' The workers haul the bale off the compressor and onto a trolley. 'This is Africa A grade, which is where you're getting most of your money from,' he says, patting a pile as we pass. Africa B grade is second, then 'Pakistan grade'. 'Pakistan grade is pure loss. The cost of handling it is more than the value,' he says. But shipping it abroad is cheaper than disposal in the UK, and at least the material will be reused.

The men shuffle the bale onto the scales. 'I put in 55kg as material weight, so my bales actually come to 56kg,' he says. Not all traders are as honest. Under-packing the bales is a common practice among smaller traders. Across the length of a shipping container, those few kilos might save the sender half a tonne in shipping costs; the customer, meanwhile, won't find out until the clothing is halfway across the world, and too expensive to send back. 'It's ripping the customer off,' Oleksii says, irritated. The second-hand business is full of such stories: traders topping up A grade bales with B material to make up for lost weight, for example, or even packing bales with rocks. 'You're not working with angels and saints here,' Oleksii says. Tales abound of African importers buying bales on credit, only to find that they've bought bales of worthless garbage. 'There are people who have committed suicide because of debt,' Oleksii says. He would never engage in such practices, he insists. 'I would rather sleep well, knowing

I've done all I can at this end.' Besides, he has a reputation to uphold.

Once the bales are ready, they – like the estimated 4 million tonnes of used clothing exported every year – will be loaded onto a container ship and set sail. For Oleksii, that's the end of the story. But for the clothing, it's just the start of a longer journey.

Fashion is the business of waste; its very existence is obsolescence. As someone who has written for a menswear magazine for much of my career, it doesn't bring me any pleasure to point this out, but it's true. The fundamental job of any clothing company is not to dress you, it is to make you want more clothes. Around 62 million tonnes of clothing is manufactured worldwide every year, amounting to somewhere between 80 and 150 billion garments to clothe 8 billion people.[12] The fashion industry produces between 8 and 10 per cent of all global carbon emissions, and 20 per cent of all wastewater.[13] And yet British adults only wear 44 per cent of the clothing they own.[14] According to one 2016 study, our closets contain 3.6 billion garments (worth £2.7 billion) that are not being worn.[15] Our wardrobes are going to waste.

And fashion's profligacy, despite many brands' recent marketing efforts trying to convince you otherwise, is only getting faster. Where once major clothing labels had four seasons per year, many fast-fashion brands can now run new 'drops' of capsule collections every day, year-round, enabled by underpaid labour in countries like Bangladesh and cheap Chinese cotton. Shen Lu, a professor at the University of Delaware who studies the fast-fashion industry, has found that H&M can add about 25,000 new products to its website in a single year, Zara 35,000. In the same time period the Chinese fast-fashion giant SHEIN – which in just a few years has gone from virtually nonexistent to comprising 28 per cent of the US fast-fashion market – added 1.3 million different new products.[16] It's hard not to think of Scott's 'Paper Caper', the paper dress marketed during the 1950s, at the height of the throwaway society. Although, as the recycling entrepreneur Ron Gonen writes in *The Waste-Free World*, '[a] $4.99 Sleeveless Jersey Dress sold by H&M is cheaper by about $2.50, in adjusted dollars, than Scott's Paper Caper [dress] was.'[17]

The result of our clothing consumption is a truly prodigious amount of waste. It starts from the factory floor: as much as 12 per cent of fibres are discarded before they even make production. Then there is deadstock, the clothing that brands order but can't sell.[18] In some cases, brands can over-order a single SKU (stock keeping unit) by up to 50 per cent. Throwing away just 10–15 per cent of a design order is seen as good practice.[19] Overall, it's thought that 25 per cent of all clothing made is never sold. The quantities involved can be astonishing. In 2018, for example, H&M admitted that it was hoarding unsold stock worth $4.3 billion,* most of which was due to be exported or incinerated.[20] The brand produces so much waste that a power station outside Stockholm, where H&M is based, switched from burning coal in part to burning clothing.[21] In most cases, this stock is simply destroyed – buried or burned. Often, so too are the 25–50 per cent of clothes that are returned, a rate that has only increased with the advent of online shopping. (When the clothes are cheap, landfilling or incinerating them is often less costly than paying someone to process the return.)[22] In total, 85 per cent of all textiles in the United States, for example, are landfilled or incinerated.[23]

* Although if you can't sell it, is it worth that much?

In the process of our conversion to cheaper clothing, we have largely lost many of the skills we once had – sewing, darning, cobbling – that once would have extended the lives of our clothes. Take a ripped pair of jeans to the tailor to have a seam fixed, and you'll quickly find that it is cheaper to buy a new pair from a supermarket or online. (Those skills, meanwhile, have been offshored to countries with the lowest possible minimum wages.) Today, many of our new clothes barely last longer than the trend. What we wear has become, increasingly, disposable.

Saturday in Accra, the capital of Ghana. Market day. Shoppers pack the streets of the central shopping district, the roads clogged with stalls and street hawkers. The traffic, unimpressed, inches through in uproar. Commerce! Everywhere bodies, movement, noise. Black Sherif booms from Bluetooth speakers, car horns blazing on the offbeat. Premier League football blares from an apartment balcony. Food carts sidle past bearing fish and plantain and bright green chilli peppers. 'You can buy everything here,' Yayra Agbofah, one of my two Ghanaian hosts, half-shouts back to me as he leads me through the crowd. It may not be an exaggeration. The variety is disorienting: toys, power tools, hob cookers, flooring, games consoles, footwear, luggage, carpets. (I think back to my tired old high street in Royston – this is more like it!) And brands – so many brands, every brand, all mixed together: Prada, Frozen, Real Madrid, Gucci, Samsung, the Yankees, Nike, Paw Patrol, Peppa Pig. Fake, authentic, 'authentic fakes'. Globalisation gumbo. In this particular market, Makola, the goods are mostly new Chinese-made imports. But we are on the trail of the used and discarded, and so we press on. When you're looking for second-hand clothes in Ghana, there is only one destination: Kantamanto.

Kantamanto is the largest second-hand clothes market in Ghana, and perhaps in West Africa. Here, an estimated 30,000 traders are crammed into just seven claustrophobic acres in the heart of the city. According to the OR Foundation, a Ghana-based non-profit organisation, 15 million garments move through Kantamanto every week.[24] (For context, Ghana's entire population is only around 30 million.) When clothes leave exporters like Pink Elephant bound for West Africa, this is where the lion's share will end up. From here, the clothes will spread across Ghana and across borders, into Côte D'Ivoire, Togo, Niger, Benin and beyond.

'Obroni!'

'Hey, Obroni!'

Foreigner! White person! The traders laugh and gossip, offering greetings and high fives. The coronavirus pandemic has meant fewer tourists passing through Accra for a while, and so our group stands out. I am being accompanied by Sena Mpeko, a British-Ghanaian journalist helping me in Accra; we in turn are being shown around by Yayra Agbofah and Kwamena Dadzie Boison, the co-founders of The Revival, a Ghanaian fashion brand that specialises in upcycling second-hand clothing. Yayra, The Revival's creative director, is a towering, elegant man with a penchant for wide-brimmed hats and wider-legged trousers. Kwamena, the slighter and the quieter of the two, with a neat beard and a taste in rings, is the brand's head of design. Together, they are two of the most stylish men I've ever met, today both dressed head to toe in black, Yayra in one of The Revival's T-shirts which reads: GHANA UPCYCLING DEPARTMENT.

The second-hand trade has a long and complicated history in Ghana, and across Africa. Its roots lie in colonialism: during British rule, which lasted from 1821 until independence in

1957, Ghanaian workers were often made to adhere to British styles of dress, which created a demand for Western clothing over traditional Ghanaian garments.[25] The trade truly exploded in the 1980s and '90s as increasingly large and well-marketed Western charities flooded Africa with clothing, intended both as fundraising and aid. Between 1990 and 2005, global textile exports increased tenfold.[26]

When second-hand textiles first arrived in Ghana, the local population had no experience of such wastefulness. In fact, they assumed the owners of the clothes must have died, leading to the Akan phrase still marked on one of the entrances to Kantamanto: *Obroni wawu*, or 'dead white man's clothes'. (In Tanzania, second-hand clothing is similarly sometimes called *kafa ulaya*, or 'dead Europeans' clothes'.)[27] But the donations, however well intended, have done as much harm as good. Unable to compete with the flood of cheap goods into Africa, local textile manufacturing sectors collapsed. Between 1975 and 2000, the number of people working in the textile trade in Ghana fell by 75 per cent. Businesses simply couldn't compete on price with a product people were throwing away.[28]

Yayra has been shopping at Kantamanto since he was a teen. 'Growing up I wanted to look fashionable, but I am not from a rich family that could afford the kind of clothes that I wanted,' he explains. 'So I started to trade or redesign stuff that I got from my brother and my siblings. Then my brother introduced me to Kantamanto, and I fell in love with the second-hand market.' A few years ago, Yayra started to hear traders in Kantamanto complaining about the declining quality of clothing shipments. He also saw it himself. 'I used to collect vintage,' Yayra explains. Once upon a time, you could find gems among the endless reams of GAP hoodies and Next jeans: Alexander

McQueen, Vivienne Westwood. Luxury fashion houses habitually slash unsold items, known as deadstock, so that it has no resale value. But sometimes uncut stock would find its way into the bales, providing an irregular supply of designer clothing to Accra's eager fashion scene. In the last few years, however, the rising popularity of thrifting and resale apps has ensured that the highest-end clothing (and its resale value) is increasingly staying in the Global North, while fast fashion has unleashed a wave of ever-lower-quality clothing on Kantamanto.

The market runs to a timetable. On Mondays and Thursdays, containers arrive fresh from the port of Tema laden with new bales. The importers and textile dealers then sell the bales on to the market sellers. 'The prices range from about $75 to about $500, based on where it's coming from, and also the grade,' Yayra says. British bales command the highest price; this is partly due to better sorting, and the increased chance of finding unworn deadstock, which sells at a markup. 'What comes in from America and Canada, you have a lot more waste.'

The bales are sold by garment type – men's shoes, women's tops – but the specific contents are a mystery, so after buying each bale, the traders will go through, valuing each item. The most prized items are 'first selection', and typically placed on hangers or displayed, pride of place. Second selection will be hung lower down. Third selection, which may be stained or faulty, is usually bundled together for people to rummage through. The rest is *asei*, trash, and if unsold at the end of the day will be swept away as rubbish. Whenever a trader buys a bale, they are making a bet that the quality of the first and second selection will be high enough to cover the losses of those clothes that cannot be sold. 'It's a game of luck,' Yayra says – one that more and more traders are losing. When the sellers can't make their money back, many get into debt with the dealers. Over time, as the quality has fallen, some have found themselves in a debt spiral, unable to get out.

Saturday is the busiest day of the week. It's today that most traders open their bales for new shoppers, who can arrive before daylight in search of the best bargains. We, however, arrive mid-afternoon, hoping that the traders might have more time to talk now the crowd has thinned. The market itself is a maze of narrow lanes, held up by simple wooden struts and a tin roof. But its simplicity hides an entire self-contained neighbourhood. Beyond the stalls, there are seamstresses; cobblers; dyers, who with a quick soak can restore a fading T-shirt or pair of jeans; a whole crew of men wielding flat irons (*cast iron* irons, heated over hot coals) to whom the sellers pay 50 pesewas (about 5p) an item to spruce up their first selection. After hours, there are barbers and food sellers and secret bars playing uptempo beats, which throng with life when work is done. We wind our way down aisles filled with racks of clothes: Asos, Dorothy Perkins, Tesco, some still with their charity shop labels on. The stalls themselves are tiny, as little as two square metres. The floor and gutters are carpeted with clothing.

Yayra picks up a ripped blouse. 'New Look,' he says, showing me the label. From England. Alongside it are blouses and skirts from F&F and Primark, one with a footprint still visible. 'The usual. Cheap fast fashion.'

Young women pass with clothing bales balanced on their head. They are *kayayei* (literally translated, 'she who carries the burden'), porters employed by the sellers to move bales around the market. The work itself is brutal: bales can weigh 55kg and more, causing spinal damage and, in some extreme cases, death. The *kayayei*, often illiterate teenage immigrants, are paid almost nothing; many live in the informal settlement of Old Fadama, a short walk from the market.

Yayra and Kwamena have been shopping at Kantamanto for so long that they seem to know everyone. Traders holler in delight as they arrive, offering warm greetings and hugs. The majority of sellers in Kantamanto are women, and so Yayra and Kwamena respectfully call them 'Auntie'. We stop by the stall of Auntie Janet, full name Janet Oforiwaa, a woman with a bright pink blouse, short dark hair and a welcoming, gap-toothed smile. Janet sells winter clothes: parkas, coats, tweed jackets. These might seem unlikely sellers in the heat of Accra, but have their own audience: fishermen, travellers and people in neighbouring Burkina Faso, where the desert nights can be as cold as the days are hot. Janet has been working in Kantamanto for thirty years, since she was a girl working on her mother's stall. How's business, I ask? 'You win some, you lose some,' Janet says in Akan, which Yayra helpfully translates.

Like Yayra, Janet has seen the quality in Kantamanto decline, felt it in her fingertips and in her bones. 'Five or six years ago it used to be good. Now, it's really bad.' The situation was made worse in 2020, when a major fire broke out in the market. Eight hundred stalls were destroyed, along with the traders' existing stocks and equipment. In places you can still see scorch marks on the roofline and along the concrete. Many sellers lost everything; Janet lost her entire stock, worth hundreds of dollars – a fortune for her. Since then, many of the stalls have been rebuilt, and there have been many fundraising efforts purportedly to help the market get back on its feet. 'But we never see the money,' Janet says. There is a rumour, which I heard everywhere in Accra, that the fire was started intentionally by real estate agents, hungry for Kantamanto's prime central location in the city.[29] The whispers seemed to be caught up in a wider sense of fear that Kantamanto's way of life is under threat.

In an outside stall we meet Vida Oppong, a denim dealer

with short red hair and a brash, outspoken manner. Vida mostly buys denim from the US or Canada, as she cannot afford bales from the UK, which for denim can stretch to $875 (around £750) a pop. I ask what's in fashion. 'Bells [bell bottoms],' she says. 'Those who like fashion buy bells. Those who don't like fashion buy the skinny jeans.' There are some customers, she says, who come to the market to buy something every week. Some will request brands by name: fast fashion, Ghana edition. I ask what's popular. 'Zara, Topman.'

'When I was a child, Next was the top,' Yayra says.

Vida has also seen the quality of imports declining since she was a child growing up in the market. 'You open the bale, and they've put *bola* [waste] inside. Plenty. I have to throw it away,' she says. Once, Vida found that an exporter had wrapped a large rock inside, hidden within the clothing. 'You want to see it? I kept it,' she says. She shows me the culprit, a grey lump of rock not much bigger than a baseball but as heavy as marble. Tipping the scales – a fraud as old as scales themselves. With margins for many traders razor thin, such a simple scam could be the difference between profit and personal ruin.

According to research by the OR Foundation, as much as 40 per cent of the clothing arriving at Kantamanto – or 6 million garments per week – immediately becomes waste. At the end of the day private garbage collectors, known as '*bola* boys', will pass through the aisles pulling carts, taking away unsold items. But collection itself costs money, and so some traders don't bother, instead leaving waste to accumulate in the aisle and the gutters. The waste is stunning; by late afternoon, there must be hundreds of garments underfoot. On our way out, through the footwear section of the market, I catch sight of a young Ghanaian shopping for sneakers in a T-shirt, no doubt imported, which says:

Overconsumption is wrecking the planet
BLACK FRIDAY
Don't buy stuff you don't need

The offices of Accra's municipal waste management department are a rundown cluster of one-storey buildings centred
around a dirt yard. Outside, collection trucks come and go,
trailing the now-familiar stench of hot rubbish. The head of
the department is a colourfully dressed and passionate man
named Solomon Noi. When I arrive in his office, Solomon is
still seated behind an imposing desk, looked down upon by a
photograph of Ghana's president. At first, he is not in the mood
to talk. Over time, Solomon has grown sick of giving quotes to
foreign journalists that translate into little action. 'I'm angry,'
Solomon says. 'We are being used as a dumping ground for the
white man's textile waste.'

As head of the municipal waste collection, Solomon sees first-
hand the impact of the textile waste on the city of Accra and its
surrounds. 'They hide under the guise of, "Oh, we are giving
donations. We are donating to charity,"' he says. 'Meanwhile,
40 per cent of the waste is complete chaff. Underwear stained
with blood. Clothing from the theatres of hospitals. Who is
coming to buy this?' Several times a week, the city's collection
trucks pick up countless tonnes of leftover textiles dumped
in the aisles and gutters around Kantamanto. Previously, the
waste was hauled to an engineered landfill in Kpone, outside
the city. But the massive influx of textile waste in recent years
created impossible conditions within the landfill, Solomon
explains. 'The textile waste soaks up water, mixing with the
dirt and the silt, and binding them together like concrete,' he
says. As a result, the landfill's compactor crews were having
to make three times as many passes to crush the waste down.

'We don't have the fuel to be wasting like that, so it means you leave it,' he says. The consequences have been stark. At Kpone, 'the void space that should take thirty to forty years to fill, was full in less than three years.'

The story gets worse. As it soaks up water, the waste clothing prevents leachate from draining to the bottom of the landfill cells. Instead, it stays still, preventing methane gas below from filtering up and escaping from the ventilation pipes. 'It is like a reactor. You are piling up gas which cannot escape.' Over time the pressure builds, until it's too late. In August 2019, the landfill in Kpone exploded, igniting a fire that burned for eight months. 'It rendered everything totally useless,' Solomon says. 'We had to cap it.' That is, bury it under a mound of earth. There's righteous anger in Solomon's voice; at times, it seems he hardly wants to look at me, preferring to answer my questions to Sena, who as a local, understands better what he is going through.

The Kpone landfill, paid for in part with a loan from the World Bank (and therefore still being paid off by Ghanaian taxpayers), was the only engineered landfill in the city. As a result of the closure, Kantamanto's waste is now being trucked to a new location, miles outside Accra, in the eastern region. 'These are not engineered landfills. They are dumpsites,' Solomon says. 'So the leachate pollutes the surface rivers and groundwater, and it is also impacting hugely on the headwaters for our water treatment company.' Water pollution from both municipal waste and gold mines upriver are polluting Accra's drinking water at an alarming rate, he says, 'to the extent that we cannot confidently turn on our taps and fetch water to drink. That is how come we've resorted to sachet water and bottled water.' The plastic packaging in turn clogs the city's gutters, causing them to overflow, and creating yet another waste crisis. Some of the clothing from Kantamanto is discarded on the beaches and washes out to sea, getting caught in fishing nets. All this can be traced back to the used clothing trade.

As the sole senior sanitary engineer in the Ghanaian government, Solomon has spent years asking for more money and attention to increase waste collection, to no avail. Meanwhile, multi-billion-dollar fast-fashion brands are making 'zero waste to landfill' pledges and paying money to Extended Producer Responsibility (EPR) schemes, which fund waste collection and treatment in the Global North. 'But *this* is where their produce has its end of life. Here. So who is supposed to get that Extended Producer Responsibility money?' By now, Solomon is nearly in tears. 'The little money we have to manage our waste, we are using that little money and space to take care of your textile waste. Which is *not right*.'

From a waste management perspective, Solomon says, the best response would be for Ghana to ban second-hand imports entirely. However, politicians have historically tolerated the import business, due to the jobs it creates – and the tax revenue it generates. Besides, even if Ghana chose to pass an import ban, it may not be able to. In 2016 the East African Community (Kenya, Uganda, Tanzania and Rwanda) announced their intention to ban the import of second-hand clothing, in order to help revive their own textile industries. In retaliation, the US threatened trade sanctions, claiming the move would threaten American jobs. All but Rwanda eventually backed down.[30]

The new municipal dumpsite is more than an hour's drive from the city, and run by a private operator that is unwelcoming to outsiders. But we don't have to go that far to see the impact of textile waste in Accra. One morning, Yayra and Kwamena drive Sena and I over to the edge of Old Fadama, on the edge of the Odaw, the river that runs through Accra and widens into the Korle lagoon. Kwamena parks us on the east of the settlement. To get in, we need to cross the river via the river barrage. Constructed at great expense in 2008, the weir was designed to intercept the city's waste from entering the ocean; a massive stainless-steel screw is supposed to lift waste from the river channel and into a waste transfer station on the shore. But, as with so many things in Ghana, the funds were misused, and the system never opened. It now lays idle and rusting, the water green and dead. Over time the weir has instead acted as a dam, against which a vast and sickening mass of plastic waste and silt has built up, so that you can now walk on the water. The waste extends almost as far as I can see, a bleak and lifeless peninsula of trash: plastic bags and packaging, electronics casings, piping, toys, clothing. The weir itself has been nearly consumed by it. A tide rising not from the sea, but from the people on the land.

Across the weir, we reach our destination: Old Fadama's dumpsite, a 30ft mound of garbage on the edge of the lagoon. We decide to climb to the top. Yayra covers his face with a handkerchief; I pull my coronavirus mask over my face to help with the merciless stench. The rubbish crackles and gives way beneath my feet as we climb. Polystyrene chunks, plastic bags, whole chunks of an old LG television, smashed eggs being picked at by flies – and underneath it all, ribbons and ribbons of clothing. Yayra and I comb through the trash, picking out labels: Zara jeans, Adidas sandals, a blazer by Polo University Club, a now-defunct Ralph Lauren brand. 'Some days you come and see fresh piles of clothes,' Yayra says.

I'll say this for dumpsites: they are at least honest. Wading ankle-deep through open dumps in the Global South might assault the senses and provide shocking pictures, until you realise that we're creating the same waste (actually, far *more* of it) at home, it's just hidden from view – whisked away to landfill, or sent away only to end up in places like this. Waste is monstrous to look at because it is a mirror. The magician shows us the rabbit at the end of the trick and we act surprised, despite always knowing that it was there from the beginning. The truest deceptions are the ones we play on ourselves.

We wade to the summit, trying not to fall. From there, we can look out on Old Fadama. At the top of the mound a herd of gaunt and sickly-looking cattle are grazing on the garbage. Longhorns. One has a tattered clothing sack tangled in her horn. She looks at me, the bag flapping in the breeze like a white flag.

The No More Fast Fashion Lab is a bright and airy space above a strip mall in downtown Accra. There are art pieces and swatch fabrics all around. The notes from a recent brain-storm are scrawled on an office wall. The lab is run by the OR Foundation, a charity set up in 2011 by Liz Ricketts and Branson Skinner, two American expats who now live in Ghana. Since its launch, the OR Foundation has been one of the loudest voices calling for reform of the global second-hand garment trade that is creating the waste crisis in Kantamanto. (If you've read or seen reports on fashion's waste problem before, chances are Liz and Branson were involved.) Few organisations have done more to map the flow of international textile waste into Ghana, or to understand the material flows within Kantamanto itself – and the human costs that come with it.

On this particular afternoon, Liz and Branson have just returned from the north of Ghana, where the OR Foundation is working on schemes to support and retrain *kayayei*, the women who haul Kantamanto's bales. Among its schemes is work with a local chiropractor, to examine of the damage done by carrying textiles on their heads. 'We knew that girls were dying because their necks were breaking under the weight of the bales,' Liz says. 'It's quite sobering. We can see how quickly the deterioration starts. Within two months of carrying, there's permanent deterioration to their spine, and they start to lose cartilage in their necks.'

Liz is a slight woman with dark hair loosely tied back, her skin freckled in the sun. She looks tired. 'I've been in a funk for the last two weeks,' she says. 'I am struggling with my mental health and my physical health because of this work.' Despite the flurries of media interest, the OR Foundation has been struggling to convert that into enough fundraising to pay its own staff or fund its programmes. 'There isn't money, outside of greenwashing,' she says. Since the OR Foundation started reporting on the impact of second-hand clothing waste in Ghana, the fashion industry has either pushed back, or sought to exploit their work for marketing purposes. 'We've had trade associations write press releases discrediting us and people we work with,' Liz says. 'Or we get recyclers and fast-fashion brands contacting us, wanting us to collect the waste here, sort it, clean and dry it, and ship it back to them for 10 cents a pound – which we could never do without slave labour, and would never want to do.' Like Solomon Noi, Liz is angry. 'We have not received one email, ever, from anyone interested in helping us clean up the current disposal sites or remediate anything.'

Since the first major global stories about Kantamanto's waste crisis broke, I've found it fascinating to watch how fashion brands have reacted. Their first instinct was to deploy the old favourite of the plastic industry: recycling. More and more clothing brands have recently started to incorporate recycled polyester and other plastic-based textiles into their products, advertising it as more 'sustainable' than alternatives. (Another reason for the pivot, which you won't see in the ads: it reduces the use of exploitative cotton from Xinjiang.) In actuality, clothing made from blends of recycled plastics and organic fabrics — cotton and polyester, say — are typically *less* recyclable than those made from a single fibre; while a PET bottle can be easily recycled into a T-shirt, a T-shirt cannot be recycled back into another T-shirt, or even a bottle. Where the technology does exist, it does not exist at anything approaching scale, and there is little to no collection infrastructure for clothing outside of the charity shop industry. Globally, only 1 per cent of clothing is recycled, according to the Ellen MacArthur Foundation; of that, only 13 per cent is recycled into new clothing.[31] Instead they are downcycled, meaning that recycled synthetics perpetuate demand for virgin plastics.[32]

Similarly, many organic textiles, such as cotton, wool and rayon, have historically been extremely difficult to recycle. That is slowly starting to change. Several companies, including the likes of US-based Evrnu and UK-based Worn Again, have gained significant investment and media coverage for developing new technologies capable of recycling clothing back into reusable fibre. At the moment, most of those are still small scale, and the details unclear. But it's hard not to look at the sheer volume of clothing being wasted globally and see in the fashion industry's recycling drive echoes of the plastics industry in decades past.

Then there are online reuse and resale apps: Vinted, Depop, thredUP, Vestiaire, The RealReal, and many others. In the US, the second-hand market is growing sixteen times faster than the sales of new clothing.[33] (It is still far smaller, both in terms of garments sold and revenue.) The trend is particularly stark in high-end and designer clothing – where the temptation to earn back a little of what you spent often trumps the altruism that leads to the charity shop.

Hannah and I have recently started to buy and resell a lot of our kids' clothes second-hand, as a way to both save money and lower our environmental footprint. It's amazingly easy, addictive, and it feels good. But by recirculating the highest quality and highest value goods in the UK, that value that we would have exported – and which traders in places like Kantamanto have relied on to survive – falls, at the same time as our shipments of poor-quality fast fashion are growing. Does that mean we shouldn't resell our stuff and try to extend the life of things wherever possible? Honestly, I'm not sure. As so often, the answer is complex, the ethics unclear. But I do believe that it's important that we recognise that our decisions about waste can have unseen consequences for people and places thousands of miles away.

Despite the work of the OR Foundation, Liz is no longer optimistic about the waste crisis in Ghana. 'We believe we have two years before one of a few scenarios happens: one is that Kantamanto is bulldozed. Two is that there's a ban in place for the second-hand clothing trade here. Three is that the fast-fashion industry sets up manufacturing here, which is likely to happen within the next year. And so you either have Kantamanto not existing, and therefore the complete loss of this model of reuse, or clothing made in Ghana by fast-fashion brands that probably went around the world within six months and ended up back here.'

The challenge for NGOs in such situations is: do you attempt to oppose the industry from the outside and force change, or do you ultimately join it, and hope to effect change from within? In the OR Foundation's case, it has chosen the latter: in

2022, a few months after my visit, Liz and Branson announced the charity had signed a deal with SHEIN to run a five-year, $50 million Extended Producer Responsibility (EPR) fund in Ghana, focused on textile waste. Some in the fashion community reacted with outrage, seeing the deal as an endorsement of SHEIN's allegedly exploitative and wasteful business practices. But for Liz, the fund was an opportunity to help the traders and *kayayei* at Kantamanto and make meaningful change, funding direct grants to traders, educational training, and new waste-disposal schemes at the market. (Solomon Noi is involved.) 'I understand people's shock, and even some of the rejection of it,' Liz told me over the phone, after the deal. 'But I'm grown up enough to recognise that we've been an organisation for twelve years. The tactics that we had been using are not working. Something needed to change.'

After I found out about the OR Foundation's SHEIN deal, I was conflicted. I think it's great that Kantamanto, and Accra, will finally see some money from the fast-fashion industry. But it also seems to me to be a sticking plaster, as well as an extremely convenient piece of PR for the fashion brands. There are many other countries that are receiving our second-hand waste – in Chile, for example, where clothing waste is overflowing from landfills in the Atacama desert.[34] Where's their money? Shouldn't something so fundamental as waste management be decided through legitimate means – taxes, international agreements – rather than one-off stunts? I can't help but think of the stuffed landfill, now capped over, and the Ghanaians who are still quite literally paying the price.

Why do we donate our unwanted things? Is it altruism, or is it the assuaging of guilt? (It can be both.) The story of aid is all too often the story of unintended consequences. Charities donate clothing to the Global South and in so doing, eliminate good manufacturing jobs. We give our things away to avoid them ending up in landfill, and unwittingly they end up on a worse dumpsite thousands of miles away. This is the consequence of our globalised waste system – it is not a flaw but how the system is currently designed. It is not so much waste colonialism, as just old-fashioned colonialism: the co-option of others' lands and livelihoods for our personal gain. We have just closed our eyes to it.

Again – it's not that I think we shouldn't be using charity shops, which are set up and run in good faith by people trying to do good things in the world. But we should also be honest, as a society, about who they are for, and the purpose they serve. Let's admit that bag of half-broken toys, or unwanted Christmas presents, isn't headed to the Salvation Army because you really believe in the mission. Let them, in turn, acknowledge that for the most part, your old television or Victoriana dining table is not going to be bought by some retro-loving local, but dismantled for parts, or burned, because nobody else wants it. And let's not let fashion brands convince us that their recycling drives are anything more than a way of ensuring that our sudden pangs of waste-conscious guilt don't impact on their bottom line. Donating isn't a salvation. For most of us, it's a simple case of making our very modern problem – having far too much stuff – someone else's.

On my last day in Accra, Yayra and Kwamena invite me down to The Revival's design studio, which is attached to Yayra's house in a quiet Accra suburb. Like them, the place is perfectly styled: buffed wood floors, music playing, the room decorated with vintage sewing machines and photographs cut from fashion magazines. The studio is a treasure trove of thrift trash. Bales of clothing are piled all around: boxy suits, swathes of stonewash denim, a box full of men's hats. On one rack Yayra has set up a little museum of uniforms: a police coat, Iraq war camo, US Navy jackets. A jacket from the US Army's 307th Signal Battalion still displays its insignia, *Optime Merenti*, 'to the best deserving'. 'We have big bags of these things, with names still on them,' Yayra says.

Every piece of used clothing tells a story of distance and time. An old leather American football helmet. A Pittsburgh Steelers jacket. A baseball cap from Mount Robson, 'Highest peak in the Canadian Rockies.' A whole rack of thick leather motorbike jackets, unusable in Ghana's tropical climate. The Revival attempts to turn some of these unusable or unsellable items into stylish, desirable objects. 'Our idea is: it's here already, we cannot send it back, we don't have the power. So we might as well just turn it to something functional here,' Yayra says. The Revival works with the skilled craftspeople within Kantamanto — the seamstresses, tailors, dyers, and cobblers — to help extend the lifespan of items that would otherwise be thrown away. Yayra takes out a bright red down jacket that they have resewn into a backpack. It's an ingenious piece of design, both sustainable and surprisingly cool. 'Now we can use it, and it won't end up in a landfill,' he says.

The Revival is currently a non-profit, and each collection is small-scale and handmade. It sells its designs in pop-up shops in and around Accra. At the moment, the operation is tiny, and can account for only a fraction of the goods arriving in Kantamanto. 'We realised that there's so much waste, and that there is not enough demand for it,' he says. Their response has been to find people who face clothing shortages, and to find ways to help them with waste. For example, in Ghana more than 80,000 fruit pickers suffer cuts and bruises while harvesting fruit crops without adequate safety equipment. 'We have about 80,000 pineapple farmers in Ghana. And there is pineapple farming all over Africa and the Caribbean,' he says. 'Subsistence farmers don't have the capital to buy protective clothing, it's too expensive.' So in 2020, The Revival developed a line of agricultural protective gear from discarded denim imports, which the brand has donated to farmers around Ghana. Yayra shows me a set of overalls which have been stitched to protect arms and limbs; the fabric itself is screen printed with a pop-art pineapple design. 'We're looking at producing uniforms for oil and sanitation workers,' Yayra says. 'And we're looking at using the leather to make jackets for commercial bikers here, because a lot of them don't wear protective clothing.'

Through The Revival, Yayra and Kwamena want to reframe the conversation around second-hand clothing – to change how we value it. In contrast to simple material recycling, upcycling conserves materials (and therefore lowers emissions). It also creates and perpetuates skills of repair and reuse, generating skilled employment, something Ghana sorely needs. 'Fashion has failed us,' Yayra says. 'Right now, sustainability is a trending thing, everybody is reporting about global textile waste – it's because something has been wrong with fashion for decades. If it was right, we shouldn't have this problem.' Yayra knows that The Revival itself can make only a small difference, and is aware of the potential hypocrisy of condemning fashion practices while running a fashion brand himself. But to him, it is 'fast fashion' that needs to end. 'Everybody's idea of fashion is new stuff,' he says. He looks animated, exasperated. 'We need to have a new culture around fashion. We should push for ownership, rather than keeping up with trends.'

To Yayra, Kantamanto is not just a warning, something for the West to feel guilty about and greenwash over. Look closely, and it's a model of what the circular economy could look like: reusing what we have already made. 'We don't need new. If we are able to create a proper circular economy when it comes to clothing, we wouldn't need to produce more clothes for the next thirty years,' he says. 'There are enough clothes produced already.'

PART TWO

FOUL

6

THE CURE FOR CHOLERA

'The sewer is the conscience of the city'
—VICTOR HUGO, *Les Misérables*

This might all sound bleak, but we've solved a waste crisis before.

It's July 1858, and Victorian London is suffocating in the grip of a relentless heatwave. The British Empire is at its peak. Smoke billows from the factories and workhouses, the docklands thronged with ships arriving from India and the West Indies. Charles Dickens, at the height of his fame, has just set off on a reading tour of the country, while on the Isle of Wight another Charles is putting first pen to paper on a book called *On the Origin of Species*. London is the most populous city in the world, thriving and filthy.

In those days, the first thing a new arrival to the capital would notice was the smell. For all of the Industrial Revolution's advances in engineering and technology, sanitation in Victorian London had remained largely unchanged since the Middle Ages. Carriages clogged the thoroughfares, leaving roads caked in horse manure. Tanneries filled the air with the scent of urine-soaked hides, soap-makers with the stink of boiling bones. Despite Edwin Chadwick's sanitation reforms of the previous decade, by the 1850s there was still little in the way of public waste collection, and so here and there the waste from homes and businesses was piled into dust heaps and dung heaps, some as tall as houses, to be picked over by the rag and bone men or eaten by swine, which roamed free in the poorer neighbourhoods. But worst of all, the entire city stank of human shit.

It's unpleasant to imagine now, but prior to the invention of modern sanitation human beings had a far more intimate relationship with our own excrement. For most of human history when nature called you simply found a private spot outdoors. In the Bible, Moses advises his followers to 'carry a stick' so that 'when you have a bowel movement you can cover it up'.[1] As settlements grew, most people would have had a designated place for when nature called – a communal latrine, a chamber pot – but in the worst cases, you went where you could. A seventeenth-century account of the Palace of Versailles in France tells of how 'the passageways, corridors and courtyards are filled with urine and faecal matter'.[2] The lavish hedgerows in the palace gardens were designed, in part, to give squatters their privacy.[3] Walking down the streets of Paris, you would have had to listen out for calls of *garde à l'eau – watch for the water!* – which indicated a chamber pot being emptied into the street from a window above. (The English mangled this into *gardy-loo*, which gives us *loo*, the nickname for toilet.)[4]

By the eighteenth century most buildings in Britain had a privy, a simple hole, sometimes with a seat, which emptied into a cesspit in the ground. Most of the time the privy would be in an outhouse, but in the cramped and overcrowded metropolis of London, privies were usually indoors, and emptied into a cesspit in the cellar. These cesspits were designed to be porous, so that the contents would leach into the surrounding soil faster than the inhabitants could fill it up. But in reality, they would frequently overflow. As Samuel Pepys wrote in his diary in 1660, 'Going down to my cellar, I put my foot in a great heap of turds, by which I find that Mr Turner's house of office is full.'[5]

When your cesspit was full you called in the nightsoil men, tradesmen – once called raykers or gong-fermors (literally 'gunge-farmers') – who scraped out the excrement, loaded it onto the back of carts, and sold it on to farms to be used as fertiliser. In the seventeenth century, arms manufacturers also worked out how to extract saltpetre from the gunge to make gunpowder; the Spanish Armada was defeated, at least in part, with Londoners' shit.[6] Though they had different names, nightsoil collectors were ubiquitous across the world, from the *vidangeurs* of Paris to China's *yèxiāngfù*, or 'nocturnal fragrance women'. (Indeed they still exist. In India, 'manual scavenging', as it's known there, is still often delegated to those of the historically oppressed Dalit caste.)

Nightsoil was a foul trade, and could be a dangerous one. Death records from 1328 tell of one Richard the Rayker, who fell through the boards of his privy and 'drowned monstrously in his own excrement'.[7] But it was also lucrative: according to the Victorian journalist Henry Mayhew, the price of having your cesspit emptied was two days' wages for the average labourer. Unfortunately, the high cost was prohibitive for much of London's poor, and so their cesspits would frequently overflow.

The situation was only made worse in the early nineteenth century by the arrival of the flushing water closet, which was just starting to be adopted in wealthy households. Toilets were not a new invention: King Minos's palace in Knossos, Crete, had a flushing toilet as early as 1,500BC.[8] The system flushed with rainwater, which flowed into a network of underground drains so large they are thought to have inspired the myth of the Minotaur's labyrinth. Similar systems have been found from the Indus valley to ancient China. The Romans – those pioneers of waste – flushed their public latrines with wastewater from the city's fountains and bathhouses, which fed into a complex underground sewer system crowned by the Cloaca Maxima, a sewer so great that it had its own goddess, Cloacina.[9] But these were the exceptions. For most of history, sewers were not designed to carry waste, but to remove surface water. Water was precious, and so was waste; as cities grew, so did the supply of human fertiliser, thereby improving the yields of the surrounding farms, in an odorous – but entirely natural and organic – early circular economy.

It was in 1775 that a watchmaker named Alexander Cummings registered a patent for a flushing privy, inspired by a design installed for Queen Elizabeth I at Richmond Palace in the sixteenth century.[10] Cummings' 'water closet' was a hit, and soon copied and improved by a series of entrepreneurs who, using modern ceramic production techniques, began to mass-produce them. In 1851, an inventor called George Jennings installed flushing closets at the Great Exhibition at Crystal Palace, which were used by more than 827,000 people.[11] For the miasma-obsessed Victorians, these cleaner latrines, or 'toilets', were a sensation. But rather than solve the city's shit problem, the water closet made it dramatically worse. With few houses connected to sewers, the additional water from the new toilets quickly caused cesspits to flood. Effluence leaked into gardens and into the streets, or settled in fetid outdoor pools. Some water closets were connected to the sewers, which had existed in various forms since the Middle Ages, but were rarely more than brick channels – some still in the open air – that drained into the city's many (now mostly underground) rivers, including the Fleet, the Effra, and the Tyburn. These in turn flowed into the Thames, carrying with them an unholy mixture of shit, food scraps, factory run-off, butchers' offal and, occasionally, human corpses.

Beginning in 1844, at the urging of Edwin Chadwick and the sanitarians, the British government passed a series of measures to try to clean up the city. Chadwick's war on garbage had been just the start. Now, he set his sights on the sewers, which he believed were among the most grievous causes of the city's health problems. Under the new drainage reforms, all cesspits were legally required to be connected to the existing sewer system, which would in turn be flushed more regularly, to prevent the build-up of miasmic gases. 'All smell is disease,' Chadwick wrote, and few would have disagreed with him.

It was a deadly mistake. Rather than solve the excrement problem, Chadwick's solution channelled the deluge of shit into the river. As a result, by the 1850s the Thames was unimaginably polluted. 'Through the heart of the town a deadly sewer ebbed and flowed, in the place of a fine fresh river,' Dickens wrote in *Little Dorrit*.[12] In 1855, the scientist Michael Faraday performed a series of experiments on the river; in one, he dropped pieces of white paper into the water, to test its clarity. After sinking just an inch, he wrote in *The Times*, the paper completely disappeared, and 'near the bridges the feculence rolled up in clouds so dense that they were visible at the surface.' Just a few years earlier, native salmon swam in the clear waters. Now it was entirely devoid of life. The magazine *Punch* regularly ran cartoons which depicted the Thames as Death itself.

It wasn't that far from the truth. Despite the putrid state of the river, many of the city's poorest inhabitants were still sourcing their drinking water from communal street pumps, which were drawn from the Thames. They couldn't then know what we know now: that a single gram of human faeces can contain millions of bacteria and viruses, along with worm eggs and assorted parasites.[13] Shit-contaminated water is the vector for some of our deadliest diseases, which inevitably ran rampant – typhoid, dysentery, and worst of all, cholera. The disease that would come to be known as the Victorian Plague arrived on British shores from India in 1831 and proceeded to tear through the populace, killing more than 70,000 people in a series of pandemics. Cholera is a disease of waste: the bacillus that causes it, *Vibrio cholerae*, infects its victim through water contaminated with faecal matter. After ingestion the bacterium makes its way into the small intestine, where it reproduces uncontrollably, triggering violent diarrhoea, vomiting and the sloughing off of the intestinal wall. The result is catastrophic dehydration. Drawings from the time depict cholera victims as gaunt, with blue skin and hollowed-out eyes. In extreme cases, victims could (and still) die within a matter of hours. In the cramped and fetid quarters of 1850s London, entire families were wiped out in a week.

Inevitably, the scientific establishment blamed miasma, and ignored any evidence that the Thames might be poisoning the city. In 1848, the noted physician John Snow traced a cholera outbreak in Soho that had claimed more than 600 lives back to a water pump on Broad Street. The well it drew from, it later emerged, had been contaminated by a leaking cesspool. On Snow's urging, the local parish commission removed the pump's handle; with the local people unable to draw water from the well, the outbreak soon ceased. Snow published his findings in an 1849 essay, *On the Mode of Communication of Cholera*, but he was summarily dismissed – and in some case ridiculed – and his theory of cholera as a water-borne disease was not widely accepted until nearly two decades later. (He is now recognised as the father of modern epidemiology.) Snow spent his final years trying to convince his peers that a clean water supply could prevent future outbreaks, but few would listen, and Snow died on 10 June 1858. By then, the heatwave had already begun.

That summer, London boiled. An unseasonably dry spring stretched into a summer-long drought. In mid-June temperatures reached 34°C in the shade, setting a new record in the city, and by early July the rivers feeding the Thames had slowed to a trickle. As the water level of the great river fell, it left its unholy contents smeared up the banks behind it – a treacle-black ooze of decaying excrement, pocked with animal carcasses, garbage, and all manner of unidentifiable filth. The stench carried for miles. It was so overpowering that people vomited and fainted in the streets; in Westminster, on the banks of the river, MPs in the House of Commons covered their mouths with handkerchiefs as they argued whether the Commons should be temporarily moved to Oxford or St Albans until the stench had abated.[14] Parliament ordered the curtains doused in lime chloride in an attempt to block the smell, and more than 200 tonnes of the chemical was poured into the river itself, but to little effect. Tell of the Great Stink, as the newspapers called it, quickly spread across the Empire. The city was in uproar. The Victorians believed, after all, that foul air caused disease, including cholera, and so the population feared for its life. Those who could, fled for their country residences, but for the millions of urban poor, there was no choice but to hold their noses and pray.

At first, the government tried to distance itself from the problem. The Metropolitan Board of Works, the body by then in charge of the city's sewers, argued that the river itself was beyond its jurisdiction. But as the drought lingered and the stench only intensified, the pressure on the government became too great. On 15 July, the Chancellor of the Exchequer (and later Prime Minister) Benjamin Disraeli stood up in the Commons to table a new bill. 'That noble river [the Thames], so long the pride and joy of Englishmen,' Disraeli said, had become 'a Stygian pool, reeking with ineffable and intolerable horrors.'[15] He urged MPs to back radical new legislation empowering the Metropolitan Board of Works to do whatever was necessary to clean up the Thames. 'The public health is at stake,' he warned. 'Almost all living things that existed in the waters of the Thames have disappeared or been destroyed; a very natural fear has arisen that living beings upon its banks may share the same fate.'[16]

As it happened, the solution was already underway.

Two summers earlier, Joseph Bazalgette, the chief engineer of the newly formed Metropolitan Board of Works, had submitted a plan to totally reshape the city's sewer system. A short, intense man with mutton chops and a thick moustache, Bazalgette was perfectly suited to the task. He was born in Enfield, north London, the grandson of French immigrants, and learned his craft as an apprentice engineer working on civil drainage projects in Northern Ireland. Before joining the Board of Works (then known as the Commission of Sewers) as an assistant surveyor in 1849, Bazalgette had worked on the railways and earned a reputation for his forensic mind and relentless work ethic. When he was promoted to chief engineer in 1856, his references for the job included the rail titan Robert Stephenson and Isambard Kingdom Brunel – two men who, alongside Bazalgette, would later be recognised as among the great British engineers of the century.

Bazalgette's plan was simple in theory, but astonishing in scope. He proposed laying some 720km of new main sewers through the city, which would connect up the existing local sewers. These, in turn, would feed into six giant intercepting sewers: three north of the river and two to the south, totalling more than 160km in length. On the north side of the river, the intercepting sewers would carry the foul water by gravity to Abbey Mills, near West Ham. There, huge beam engines would pump the sewage out even further, to Beckton in Essex, where it would be discharged into the Thames at high tide. On the southern banks, the sewage would be transported to Crossness in Kent,[17] where another pumping station would pump it up into outfall reservoirs and ultimately into the North Sea.[18]

The plan was a breathtaking feat of imagination. To even begin, Bazalgette's team had to map the entirety of London above and below ground, taking in not only elevation and geology, but also the convoluted existing networks of sewers and underground rivers, some already centuries old. It would involve digging up huge swathes of the city, including major thoroughfares: to build the Piccadilly branch, Bazalgette's engineers dug a huge trench down the centre of one of London's busiest commercial streets. Two railways had to be lowered, and several roads raised. But the most remarkable feat involved the river itself. To lay the intercepting sewer along the north and south banks, Bazalgette designed three enormous embankments: the Victoria and the Chelsea embankments to the north, and the Albert to the south. These structures, which today contain both the intercepting sewers and the Circle Underground line, totally reshaped the shoreline of the Thames. The Houses of Parliament, which had opened onto the water for centuries, were now set back from the river, and a new road bisected the city, connecting Westminster in the west with the Bank of England further east.

For two years, Bazalgette's plan had been caught up in a bureaucratic standoff. Now, with the Stink still lingering, Parliament passed Disraeli's emergency bill, empowering Bazalgette to proceed. Construction began in January 1859 and continued for another fifteen years. By the time it was finished, Bazalgette's engineers had used more than 318 million bricks and laid more than 670,000 cubic metres of concrete.[19] Among the many innovations made in the process was the development of Portland cement, the basic ingredient for modern concrete, and quite literally the foundations of the modern world; today, concrete is – besides water – the second most widely used substance on the planet, thought to be responsible for as much as 8 per cent of human CO_2 emissions. In total, Bazalgette's sewer system cost £6.6 million, or around £825 million in today's money. It would prove a bargain.

It might seem strange to us that in Victorian England sewers were considered exciting, even beautiful. In 1865, the Crossness pumping station was opened by the Prince of Wales, and celebrated with a grand banquet. The pumping stations themselves, which still stand today, are remarkable feats of utopian design: tall, sunlit buildings, laced with intricate wrought-iron metalwork, they have the feel of gothic cathedrals rather than buildings designed to pump faeces. *The Observer* described Bazalgette's feat as 'the most extensive and wonderful work of modern times'.[20] Similar excitement followed elsewhere. In Paris, the sewer system designed by Georges-Eugène Haussmann and Eugène Belgrand was a popular tourist destination; gentlemen in top hats and ladies in crinoline dresses were conveyed along the lamplit sewers in *bateaux* to admire the underground architecture. (Unlike London, Paris' system did not at first carry excrement; when that was resolved decades later, the tours became less popular, though you can still see some of the original tunnels at the city's Sewer Museum.)

London was not alone. Beginning in the 1850s and for the rest of the century, cities around the world were remade by equally grand sewer-building projects. In Chicago, for example, the city traced a series of cholera outbreaks to drinking water from Lake Michigan, where its sewage was also being dumped. To solve the problem, engineers raised entire streets. Workmen dug underneath the wooden buildings, placing them on hundreds of jackscrews, which were then cranked up inch by inch while new foundations were laid underneath.[21] Other buildings were lifted onto log rollers and moved wholesale. Eventually, in 1900, the city's engineers physically reversed the course of the Chicago river so that sewage would drain southwards, through a series of artificial canals, into the Mississippi basin. It was a herculean feat of civil engineering, one that remains controversial to this day; although it solved Chicago's sewage problem, it created a series of new ones for the communities and habitats now downstream. Invariably, the cities that were slowest to solve their waste problem by building sewers were also the ones who continued to be plagued by disease.

In 1866, London suffered its final cholera epidemic. The outbreak, centred around Whitechapel in the east of the City, took 5,596 lives.[22] As it happened, Whitechapel was the only district not yet connected to Bazalgette's sewers, which were finished the following year. The Whitechapel tragedy was vindication of London's sewer project, and a posthumous redemption for John Snow, whose theory of cholera as a water-borne disease was finally accepted by the medical establishment. In 2018, the London School of Hygiene & Tropical Medicine and Westminster Council installed a memorial pump at the site of the Broad Street outbreak. In a nod to Snow's achievement, the handle is missing.

Bazalgette's sewer system, along with the work of Chadwick and the sanitarians, did more than solve London's stench problem. It reframed the relationship between the city's inhabitants and their waste. Once a visceral feature of everyday life, excrement was suddenly removed from our public consciousness. Shit still happened, but in private. And with the advent of automobiles in the early twentieth century, faeces, something as common as mud for humans for thousands of years, all but entirely disappeared from modern life, conveyed underground and out of sight as if by witchcraft. Cholera, dysentery, typhoid and many other then-common diseases were all but eliminated by the introduction of the sewers; it's thought that their introduction in the nineteenth century added twenty years to the average human life.[23]

Of course, that was not true for everybody. Just as with solid waste, the sanitation we enjoy in modern cities is far from universal: 1.7 billion people worldwide still do not have access to modern sanitation facilities.[24] Every day, an estimated 494 million people without access to flushing toilets and closed sewers are forced to shit in the open, in gutters or in plastic bags. These people are inevitably the poorest: those living in slums or stricken by war. And despite the relative sterility of the modern city, the World Health Organization estimates that one in ten people consumes wastewater – that is, sewage – every year, either via unclean drinking water or contaminated food. Diseases of waste are still killing countless people. Every day about 2,200 children die of diarrhoea, most of them without access to reliably clean water.[25] Every year, poor sanitation kills more children than AIDS, malaria and measles combined.[26]

When Bazalgette designed London's sewer system in the

1850s, the city's population was 2.5 million. He could have built a sewer system with the capacity needed for the time, but instead he doubled it, anticipating that the city would continue to thrive and grow in the centuries to come. The Victorians built the system to last, with the egg-shaped tunnels solidly built from concrete and blue Staffordshire brick. And last they have: Bazalgette's sewers remain the backbone of London's sanitation system to this day, whisking away the city's foulness beneath its inhabitants' unsuspecting feet.

Yet despite Bazalgette's great foresight, he couldn't possibly have anticipated the scale of modern life. Nine million people now live in London,[27] each of them using many times more water than the average person of Bazalgette's era. Flushing toilets, showers and baths, bidets, washing machines, dish washers, garden hoses, and the many other water-guzzling devices that make up our daily rituals produce an immense quantity of wastewater that is now flushed down our drains. Inevitably, as London has expanded the network has been modernised. Beginning in the 1880s, a series of sewage treatment works were constructed to treat the raw waste before it is dumped into the sea, and repeated expansions have extended Bazalgette's network far beyond the initial reaches of his system. But the Victorians' utopian vision of sanitary engineering quickly subsided. Contemporary sewage treatment plants are ugly, utilitarian affairs, and perhaps the city's most vital infrastructure is built and maintained out of sight, by people we rarely meet or think about, yet upon whose labour modern life depends.

The more I learn about how Bazalgette solved the Victorians' waste crisis, the more I want to see it for myself. Which is why on a cool, clear November morning I find myself climbing out of a taxi and walking up the long gravel drive of Mogden Sewage Works in Isleworth, south-west London. Mogden is the second largest of 350 treatment works operated by Thames Water,[28] the monopoly that handles running water and sewage for the 14 million people who today draw their water from the river. Built in the 1930s, the site is vast: 140 acres of boxy pre-war buildings and water tanks, veined above and below ground with thick steel pipes. Still, unless you are looking, you would barely know it's there. The site is surrounded by trees, an anonymous island in the middle of a leafy suburb; Twickenham stadium, the home of the English rugby team, lies just beyond the outer boundary to the south.

The only thing that may give it away is the smell. On hot days the fetid scent from the works can still carry on the wind, a situation that has long caused problems with local inhabitants. In the run-up to the 2015 Rugby World Cup, the scent drifting from Mogden gave the stadium the nickname 'Stinkenham'. As a result, the site recently completed a series of upgrades to remove the stench, covering many of the formerly open-air treatment ponds and installing a series of monitoring devices to detect spikes in malodorous chemicals. On the day I arrive it's barely perceptible, an umami sourness somewhere between fresh compost and wet dog.

'I love that smell,' Dina Gillespie, Mogden's site manager, tells me. Dina's a tiny, warm woman, with rimless glasses framing a round face and a curling shock of greying black hair. We meet in her office, inside the main management building. It's clean and neat. There is fresh fruit in an enamelware bowl, an orchid on the windowsill, and pristine work boots arranged in a row below the coat pegs. Behind the desk is a white-board covered in legible, well-ordered notes about incoming water flows, energy usage and tank rotation. Since the recent upgrades, Dina explains, complaints about the smells have fallen. That doesn't mean there aren't days when it's bad: in hot and dry spells the water levels in the sewers fall, and so the solids build up and turn septic. If heavy rainfall then flushes the system through, all that stench – a Great Stink in miniature – ends up at Mogden in one go. But that's a relative rarity.

In sewer terms, Dina is a veteran. Prior to taking over Mogden, she managed Beckton sewage works, the site of Bazalgette's original outfall sewer, and now home to Europe's largest treatment plant. Beckton handles 14,000 litres of sewage a second. Mogden is older and smaller – it can handle up to 12,000 litres of sewage a second, or a billion litres a day – and so was deemed to be in greater need of her skills.

We don our protective equipment for a tour. One of the things working in sewage gives you, Dina says, is a finely tuned sense of the weather. Heavy storms, cold snaps – each present their own unique challenges, and their own smells. After enough time, she says, you can navigate by it. 'I know the differences,' she laughs. 'I can smell when it's the inlet, when it's crude sewage, when it's the aeration lanes or pasteurisation.' To Dina, sewage isn't just the mechanics of flow, of pipes and filters and pools. The plant is more like an organ, digesting everything we feed into the live organism that is the city. 'People may think it's easy, but it's a really tough job,' Dina says. 'Because it's a biological process, it can quickly get out of control. There's a number of things you have to juggle to keep it going without it tipping over the edge into disaster.'

Dina has a close personal appreciation for sewers in part because she didn't have any growing up. She was born in Sudan, and her family, which is half-Sudanese, half-Scottish, drew water from a well fed by the Nile. Eventually, she came to the UK as a teenager and studied civil engineering, before deciding to work in waste. Living in places without modern waste systems, she says, makes you appreciate the mundane miracle that is the sewage system. 'I think it's absolutely amazing,' she says. 'Yet nobody knows what we're doing. The general public today just flush the toilet and that's it – but we are so critical to the daily lives of Londoners that it's unbelievable. If we didn't do what we do, there'd be big problems.'

When the local intercepting sewer reaches Mogden, the flow first passes through a series of mechanical screens, which remove any solid objects – trash, rocks, the occasional bike – caught up in the flow. The sewage then flows into primary settlement tanks, circular ponds the size of swimming pools, where it is slowly stirred so that the solids fall to the bottom. The solids, known as sludge, are then dredged, dewatered and passed through anaerobic digesters to produce methane. Burning the methane provides 60 per cent of Mogden's energy; many more modern sewage treatment works now act as power plants, providing their surrounding neighbourhoods with renewable energy. Anything left after digestion is piped to another plant at Iver South, near Heathrow, where it's turned into fertiliser – the age-old practice of harvesting nightsoil, reborn for the twenty-first century.

With the sludge removed, the remaining water in the settlement tanks is pumped into aeration lanes – long, rectangular pools that extend across the site in stripes. Air is blown in at high pressure, encouraging aerobic bacteria to digest any remaining pathogens left in the water. The process also acts to break down ammonia, a toxic pollutant found both in faecal matter and urine. 'In crude sewage, the ammonia might be over 50 milligrams per litre. Once it's been through the aeration lanes, that's then down to less than one milligram per litre,' Dina says. Finally, after a last round of settlement in smaller tanks, the cleaned water is expelled back into the river.

Several times a day, on-site scientists test the water quality for pathogens and chemical content, to ensure they are below government-mandated safety levels. Sewage, like trash, can tell you a lot about a city's population: its eating habits, its flushing habits, and even its health. Several cities around the world monitor sewage to detect diseases, including cholera; shit will normally tell you about an emerging outbreak before hospital admissions do. During the coronavirus pandemic, London, like many cities, began testing the viral load in their sewage in real time to track cases and detect the emergence of new variants. In 2022, sewage testers detected polio virus in the city's sewage, triggering a hurried vaccination drive – shit as early-warning system.

We walk along the lanes. It's quiet. With modern sewage plants largely automated, only eighteen people work on the site at Mogden, giving the whole place an eerie beauty. Gulls and collared pigeons sweep over head, occasionally landing on the tanks, looking for scraps, and the breeze blows golden autumn leaves down into the effervescent water.

The greatest challenge facing sewage systems today is not shit, but 'rag'. Rag is the waste industry term for the thousands of tonnes of non-flushable trash that is nonetheless put down the drain every day: condoms, nappies, cotton-wool pads, tampons, and, chief among them, wet wipes. The UK uses 11 billion wet wipes every year.[29] Their thin, often-plastic fibres are the bane of modern sewage systems, causing an estimated 90 per cent of the 75,000 blockages that Thames Water has to clear out every year. Dina sighs. 'It's always the wipes.'

Inside sewer pipes, wet wipes have a tendency to snag on

surfaces or get caught in grates. Fats and oils then stick to the fibres, slowly building up in layers. At their worst, these blockages snowball into 'fatbergs', immense fat-white plugs of decomposing putrescence that block entire sewers and cause the system to burst and flood the streets above. In 2017, Thames Water engineers uncovered a fatberg in Whitechapel that was 250m long and weighed an estimated 130 tonnes. To remove them, Thames Water employs specialist teams of unblockers, who must be lowered into the tunnels in respirators and full hazmat protective gear. They break up the putrid mass with heavy tools (the mixture of fat, rag and surface grit, set hard, is apparently like mining wet rock) and remove it piece by piece, often by thickly gloved hand. It's hazardous work. The mass can shift and collapse, while sharp objects like syringes can get stuck in the mass ready to puncture any protective layers. There's a risk that toxic gases can build up behind blockages – in Bazalgette's time, methane build-up in the sewers occasionally led to explosions – and so fatberg miners work in thirty-minute shifts before returning to the surface. (I'm told, to my dismay, that journalists are no longer allowed down with the breakup crews for the same reason.) Breaking up the 2017 fatberg took eight workers more than nine weeks. When the task was done, a slice of it was put on display in the Museum of London, where visitors could watch it decompose in real time.

Dina isn't sure how much rag is removed at Mogden, but at Beckton, the filters catch 30 tonnes a day. 'It's a whole operation on its own,' she says. 'We can't remove one hundred per cent of the rag, it's just impossible, so the stuff we don't remove gets into the aeration lanes and into the pumps and digesters. We have to replace pumps because of rag. Cleaning it up costs a considerable sum of money.' The problem extends beyond the sewers. So many wet wipes and other textiles have washed into the Thames that they have formed a 'wet wipe island' in Hammersmith, big enough to alter the river's course.[30]

Rag is not the only strain upon sewer systems around the world. As London's population has grown, it has put increasing pressure on sites like Mogden. Where once rainwater would drain into the soil, today our concrete-carpeted metropolises ensure that sewers are the only place for all that water to go. And so, when the rainfall is too heavy for sewers to handle, they overflow. Mogden has eight storm overflow tanks, but they are nowhere near enough, and so raw sewage is increasingly forced through the overflow pipes and directly into the Thames. The problem is getting worse: in 2021, raw sewage was discharged into England's rivers 375,000 times, totalling more than 2.7 million hours.[31] In London alone, 39 million tonnes of discharge – billions of litres – still runs into the river every year, much of it from Mogden. (It is not a uniquely British problem. In the US alone, as many as 5.4 trillion litres of raw sewage is discharged into rivers and lakes every year.)[32] The problem is exacerbated by climate change, which is driving more extreme weather events. And it's not just the rivers: Britain's beaches are similarly being flooded with sewage at a horrifying rate. In 2022, dozens of beaches in England and Wales issued pollution warnings and were forced to close to the public after being flooded with sewage after heavy rain. Many swimmers have reported falling sick, allegedly from ingesting untreated sewage. Over the course of reporting this book, the issue has grown into a national crisis.

Raw sewage isn't just loaded with deadly pathogens. The mixture of chemicals – ammonia, nitrogen, phosphorus, and heavy metals – can trigger algal blooms that starve waters of oxygen and poison wildlife. As such, although the Thames today is still cleaner than in Bazalgette's time, swimming in the river is discouraged. (You *definitely* shouldn't drink it.) Ironically, because raw sewage is a risk to human life, many of London's sewer outfalls are today nature reserves. Follow Mogden's outfall pipe underground and it eventually emerges on Isleworth Ait, a small teardrop-shaped island in the middle of the Thames. The islet is crowned with sycamores and willow trees that droop over the banks; passing boats are known to spot kingfishers and herons fishing in the water, presumably unaware of its toxic contents.

To solve the Thames's latest waste problem, engineers are currently working on the biggest overhaul to the city's sewer system since Bazalgette's lifetime. In 2016, work began on the Tideway Tunnel, a vast new intercepting sewer running 25km from Acton in the west of the city all the way to Abbey Mills in the east, intended to connect thirty-four of the existing sewage overflow pipes – including London's old underground rivers – into one, so that under heavy rain the raw sewage no longer flows into the Thames, but into the great tunnel beneath it.

It's February 2020, and I'm standing almost in the middle of the Thames. Sun mirrors off the water's surface. Clouds dapple shadows across the glass and concrete skyline. It would be serene, if not for the incessant crunch of drills on bedrock. I've come to see the Tideway Tunnel under construction. This particular site, just downriver of Vauxhall Bridge on the south shore, is actually in the river itself. A cofferdam – a retaining wall of steel pylons driven into the riverbed – juts out into the river like a spit, holding back the water while the engineers tunnel into the bedrock below. Materials arrive and depart every few hours by barge. Of all the tunnelling sites that Thames Water could have invited me down, this seems the most fitting: the cofferdam extends out from the Albert Embankment, Bazalgette's original design, which now holds up the headquarters of MI6, one of the UK's intelligence agencies. On the north bank opposite the flags of the Tate Britain art gallery flutter in a westerly wind.

Here, below the river, the sewer outflow from the Effra – the underground river that winds up from Brixton to the south, through the Borough of Lambeth – will eventually connect to the Tideway. The Effra's existing outfall, a heavy steel gate, runs out under Vauxhall Bridge. Mercifully it's been a dry spell, and the outlet is quiet. 'When we had a huge downpour in September there was a huge discharge,' Lee Fisher, the manager of this building site and (along with a PR person from Thames Water) my guide, tells me. 'You could see the slick of fat in the river. People were complaining about the smell.'

Lee is a reassuring presence, tall with soft features and skin ruddy from working outdoors and on the water. He's a tunneller, not a sewer man: before Tideway his last job was managing the recently renovated Crossrail and Underground station at Tottenham Court Road, in the city's West End. After a short safety briefing, we head down a spiralling scaffold staircase and into the tunnel.

To descend into the Thames is to go back in time. Below the riverbed – three metres of sediment, gravel, and several centuries of waste that carpet the bottom – London sits on a bank of clay 20 metres deep. 'It's like a plug for the water,' Lee explains. 'It's a good material to mine in, clay. That's what the Underground is dug in.' Tunnelling is both construction and archaeology: early on in the tunnelling phase, Tideway engineers uncovered Mesolithic wooden beams near this site, believed to have been part of a bridge over 8,000 years ago, proving there was likely a river crossing here long before there was a human settlement. Dig even deeper, and a million years pass every few inches. 'One of the bands we came across, called the upper shelly beds, was absolutely full of shells – and those shells were laid down 50 million years ago,' Lee says.

As with Bazalgette's sewers, the Tideway tunnel works by gravity. At Acton, the main tunnel is 35 metres deep, falling to 75 metres deep at Abbey Mills. 'Part of the reason for the depth is that we just go underneath everything. All the tube lines, telecoms tunnels, power tunnels, water tunnels,' Lee says. In places, the city's hidden infrastructure extends as far below ground as the skyline does above it.

Inside the hole, bright yellow beams almost a metre thick are holding back the surrounding earth and water. At the bottom, a worker is digging away in an excavator, perched atop a mound of rubble that is waiting to be lifted away by crane. When the tunnels are connected, the raw sewage will flow out of the new connecting tunnel close to where we're standing and spiral down into the main shaft in a vortex, essentially a water slide of effluence. (The vortex prevents the sheer weight of sewage cracking the sewer lining as it falls.)

When the Tideway is finished, Thames Water claims, 95 per cent of the raw sewage that pollutes the central stretch of the river today will no longer flow into the Thames. Fish are expected to return in much greater numbers. There may even be swimming; similar cleanup schemes in Oslo harbour and parts of Paris have allowed swimmers to return to the waters, bringing the city's inhabitants closer to the natural environment around them. By then the cofferdam will be gone, but several inspection shafts will still jut into the river, terraced and planted as public parks, providing new habitats for river plants and crustaceans, as well as somewhere for people to sit and admire the cleaner waters.

The bottom of the tunnel is a surreal place to be, quietly profound. Few humans have ever stood on this spot on the Earth, deep in this ancient river. After the Tideway is finished in 2024, it's unlikely they ever will again. The Tideway is designed to last at least 120 years, with inspections every ten, 'although by then, I imagine it'll be done by drones,' Lee says. By 2060, Thames Water estimates, London will be home to 16 million people who will rely on this tunnel every day. Most of them will likely never know of the effort that went into the waste system beneath their feet.

'The best part of this job is the sunsets,' Lee says, as we climb back up the stairs and into the daytime. Just as everything smells sweeter after you've been among rubbish, the darkness below ground gives one a new appreciation for the light. We look out at the Thames. 'You can see Parliament, the London Eye, the river . . . it's beautiful.'

The Tideway Tunnel is in part the legacy of a decision Joseph Bazalgette made sitting at his desk at the Metropolitan Board of Works back in the 1850s. Combined sewers, in which rainwater and sewage are mixed together, were at that time the most practical and expedient solution to Victorian London's waste problem. And they worked: in other cities, which built rainwater-only sewers and relied on cesspits to deal with human waste until several decades later, water-borne diseases lingered longer and took many more lives. But today, sanitary engineers prefer to use a separated sewer system, where excrement is sent to treatment works but stormwater is channelled away separately into the river. Such systems are more expensive, and imperfect – our ever-growing cities produce too much waste for that – but more sympathetic to the environment, and so many new cities, particularly in Asia, are being built with separated sewer systems, in the hope of reducing sewage overflow and cutting water waste. Would everyone be better off with separated sewers? Maybe. With rising sea levels and changing weather patterns likely to inundate more and more cities in the decades to come, improvements will need to be made now so that future floods – or one long, hot drought – don't return us to the age of Great Stinks.

The Victorians' solution to their waste crisis was flawed, but it was also remarkable. Sewers, even more than railways, bridges or steam ships, may be their greatest contribution to modern life. In a 2007 poll by the *British Medical Journal*, mass sanitation – boring old toilets and sewers – beat antibiotics and painkillers to be crowned the most important advance in medical history.[33]

There are lessons to be taken from Bazalgette and the sanitarians, not all of them good. Yes, the British government of the time gave Bazalgette's Board of Works almost unlimited resources to begin building, but only after cholera and other diseases had cost untold lives. It wasn't until the fetid stench was singeing their nostrils that they finally took action. (There may be a depressing precedent here for climate change.) Bazalgette and his engineers built the sewer system without fully understanding the implications for public health. They couldn't know then the bacterial cause of their plight – just as we're only beginning to understand the effect of pollutants and microplastics on our own health and the environment. The massive short-term disruption caused by the sewers must have seemed horrendous to the city's inhabitants, but the benefits were worth the cost. Wider paved streets, new Underground train lines, and the beautiful, stone-walled embankments looking out over a cleaner river transformed London into the city we know today. Just as cities were built on trash, they were beautified by sewers.

Today, sewers are more than a mundane necessity under our feet. They're increasingly urgent. It's not just that one in four human beings is still, remarkably, without access to everyday sanitation, or the millions of people without even toilets, who defecate in plastic bags. As climate warming heats the planet, water is becoming an ever more precious resource. By 2050 the number of people worldwide facing water scarcity is expected to double.[34] This has already begun, with cities in India, Africa, and the United States running out of water, and aquifers feeding some of the world's most important rivers running dangerously low. In the UK and abroad, chronic under-investment by governments and by profiteering utilities companies have left our modern sewage systems, as well as our rivers and beaches, in a dire state. After spending time down in the tunnels, one thing is clear: we need new Joseph Bazalgettes – and fast.

7

A THIRD OF EVERYTHING

'See? They're absolutely perfect'

A joyful cry emerges from the dumpster: 'Pears!'

It's a bright spring day in York, the ancient walled city in the north of England – city of Grand Old Dukes, Wars of the Roses, and Yorkshire puddings – and I'm in a car park behind a grocery shop, rooting through the bins.

'Ah,' my guide, John Cossham, says, gleefully. He's leaning so far over the rim of a huge green dumpster that his torso is nearly submerged, scooping through handfuls of packaging for the neglected treasures beneath. There's a light stench in the air, and a thin coating of grime over everything as he fishes the fruit out and places it in the trailer on the back of his bicycle. The pears are lightly bruised and need a thorough wash but otherwise, John reckons, perfectly serviceable. 'Mmmm,' he says, turning one over appreciatively. 'Ripe pears in the rubbish is exactly what I want.' There's plenty more, too: bananas; apricots; big, juicy-looking tomatoes; pak choi; raspberries and potatoes – some blemished, most a little squashed, but all, John says, perfectly edible.

I met John online, where he runs the UK's largest Facebook group for freegans – that is, people who dine out on what others throw away. The freegan movement, which grew out of 1960s California counterculture, has become synonymous with skip-diving, but really it's something more fundamental: a spiritual rejection of consumerism and waste.

If I asked you to picture a freegan, John is more or less exactly what you might imagine. At fifty-four, he has the weathered complexion of a life lived largely outdoors, and the brisk, musky scent of a man unconcerned by hygiene or other people's judgement. (The latter, you feel, being a prerequisite for digging through bins in broad daylight.) John's T-shirt is moth-holed, his jeans stonewashed, and his toes visible through leather sandals. His long hair, greying and balding on top, is nonetheless pulled into a ponytail, which complements his beard. In sum, the aesthetic is somewhere between affable climate activist and festival drug dealer – which, combined with his authoritative knowledge of all things food waste, give him the uncanny magnetism of a prophet, or the leader of a particularly unsanitary cult.

'Oh, I'm in heaven! I've just found a load of mangoes,' John says, hauling out his prize, which are soft and gleaming green. You can blend even bruised and out-of-date mangoes into a perfectly good smoothie, he explains, or chop them up over granola for breakfast. If fruit is particularly battered, he'll dry it out – mango leather is a delicious snack, chewy and long-lasting. He loads those into the cart, too. John's enthusiasm is so infectious that I soon get over my initial disgust and find myself crawling around on the pavement, wiping smushed raspberries off potatoes that have spilled into the gutter.

I'm a little nervous. Although it's not technically illegal in the UK to take things from bins, it is prohibited on private property, and freegans have in the past been prosecuted – particularly by businesses who would rather their waste remain secret. Many large retailers now lock away their bins at night to prevent freegans gaining access. Thankfully, on this particular day we have permission from the shop's owners. John dives this bin pretty regularly, and the staff at the shop, Millie's, will often leave unsold food out for him. In addition to using the waste to feed himself and his family — he has a wife and two adult sons — John collects wasted and surplus food from around York on his bicycle and helps to distribute it around local food banks and homeless charities. Anything left over he turns into compost, which has earned him the nickname Compost John.

He is something of a local celebrity. In 2008, the charity Oxfam ran a competition to find the UK citizen with the lowest carbon footprint. John won. He doesn't drive, fly, own a television, or even a toaster. He also doesn't shower, instead washing with cups of cold water each morning. (This, I think to myself, explains the smell.) In the end, John's footprint was calculated as being less than a quarter of the average UK person. 'My lifestyle is all wrapped up in using as few resources, being as efficient and low energy as possible, so that you're creating as little waste as possible,' John says.

The dumpster is truly disgusting: fermenting cabbage skins, obliterated peaches, and other less easily identified garbage sandwiched between cardboard and plastic crate-wrap, the layered effect bringing to mind a particularly fetid *mille-feuille*. Beneath one layer we uncover several boxes of eggs, only a day past their expiry date. 'I can eat those,' he says. 'I hard boil them, put them in vinegar, and make pickled eggs.' There are yogurts, too, which can be rescued provided they're still sealed and the weather hasn't been too hot. 'I find a lot of dairy. I'll eat dairy, but I won't take any meat.' This is as much down to the health risks — raw and rotting meat can be riddled with E.coli, salmonella, and other bacteria — as any dietary beliefs. 'I'm not a vegan, because I'm a freegan,' he says. The important thing is that the food doesn't go to waste, and also that it's free. John pats his overhanging stomach paunch and grins.

'That,' he says, 'is not paid for.'

I've always felt a special kind of guilt about wasting food. My mother is a chef and a food writer, and many of my fondest childhood memories are of eating. There were always bowls that needed licking, leftovers to pick at, some freshly baked experiment cooling on a countertop. At mealtimes, Mum (like my Nana before her) was adamant that my brothers and I should clean our plates. I've had a strong aversion to throwing away food ever since. Then my daughters were born, and as I wiped up half-chewed carrot sticks from the floor and scraped browning banana splatter from the walls, mealtimes became a visceral daily reminder of just how much perfectly edible food we throw away. The numbers are stomach-turning. Worldwide, more than 931 million tonnes of food goes to waste every year, according to the United Nations Environment Programme.[1] Estimates vary – data on food waste being hard to collect even by waste's standards – but it is thought that up to a third of all food produced worldwide is discarded without being eaten.[2] That's about $1 trillion's worth every year, enough to feed 2 billion people, or every single undernourished person in the world four times over.

Food waste is not just a human tragedy, it is also an environmental one. Between production and disposal (rotting food, if you recall, throws off huge amounts of methane), all that waste generates approximately 3.3 billion tonnes of greenhouse gases. According to the Intergovernmental Panel on Climate Change, that means as much as 10 per cent of all global greenhouse gas emissions can be attributed to food waste.[3] If it were a country, food waste would be the third highest emitter on Earth, behind only China and the United States.[4]

Almost anywhere you find food, you'll find it wasted. It's wasted in the fields, thrown away during sorting, discarded during manufacture, and tossed out at retail. The majority of food waste, however, comes at home: it is the food we leave rotting in our refrigerators and scrape off unfinished plates. Our profligacy when it comes to food is not just a Western phenomenon. It's universal. China throws out 350 million tonnes of food every year,[5] India 68.8 million.[6] (That might seem a lot, until you remember both have populations in excess of 1 billion people.) According to official figures, the US discards an estimated 63.1 million tonnes of food waste[7] – putting it among the more wasteful nations per capita, which I understood viscerally the first time I encountered American portion sizes.

Experts like to differentiate between food *loss*, the food wasted on farms or in manufacturing, and food *waste,* that which is lost at or after the point of sale, by restaurants, retailers, or consumers. This is for good reason: while food is wasted everywhere, it not always for the same reasons. In the Global South, a disproportionate amount of food is lost in the production process – due to factors such as the warmer climate, inadequate storage or refrigeration, and pests – whereas in the Global North, the majority is wasted, having been thrown away within households. Differences can also be cultural. The Quran, for example, expressly condemns food waste; whereas in many parts of China, it is considered rude to finish your plate, lest you insult your host's generosity.[8] So much food is wasted by this practice that in 2020, the Chinese government passed a law that instituted fines for restaurant customers who leave excessive amounts uneaten on their plates.[9]

According to the food charity WRAP, the UK throws away

6.6 million tonnes of edible food per year.[10] That's the caloric equivalent of 10 billion meals. Those numbers can be difficult to comprehend until you're semi-submerged in a binful of it, so it might be better illustrated through a food that we discard more of than almost any other: bread. The UK alone wastes more than 900,000 tonnes of bread every year, or 20 million slices of bread *every day*. According to the supermarket chain Tesco, 44 per cent of all white sliced bread in the UK is never eaten.[11] In the USA, a third of all bread is wasted.[12] (Try picturing that, the next time you ask for your crusts to be cut off.) Most of that waste happens at home, because we buy loaves that are too big, and so go stale or mouldy before we get a chance to finish them. Then there's the bread lost in retail and manufacturing: in sandwich factories, for example, it's not uncommon for the end slices of every loaf to be thrown away (nobody wants to buy a sandwich made with the crust piece). And that's just *bread*. Forty per cent of salad leaves in the UK go uneaten. So do 400,000 tonnes of meat (chicken, the cheapest meat, is also the most wasted) and 490,000 pints of milk. The average UK household spends £700 per year on food that they don't actually eat.[13] Meanwhile, 4.2 million Britons currently experience food poverty.[14]

Crucially, food waste doesn't just mean losing the potatoes or salad leaves you're tossing out at the end of a packet. It also means that up to 1.4 billion hectares of land used for farming – as much as 28 per cent of all farmland worldwide – is being wasted, when instead it could have been put to other uses, such as housing or planting forests.[15] According to the World Wildlife Fund, the farmland we use to grow uneaten food every year could cover the entire Indian subcontinent.[16] Seventy per cent of the planet's fresh water usage is for agriculture, about 4 trillion tons of water per year in the US alone.

If we assume a third of that food is wasted, that means one in five litres of freshwater extracted globally every year is being used to grow food that is never eaten by people. (A lot of that food waste is fed to livestock, but more on that later.) In fact, the resources spent on producing wasted food are even more valuable than the food itself. What matters isn't a single harvest, but what else could have been done with all that land, that water, and labour. Consider it this way: instead of growing a tomato that ends up being thrown away, that farmer could have used that land and energy to grow wheat, which is more energy-efficient, calorically dense, and less perishable. Or, as the food waste activist and former freegan Tristram Stuart puts it in his excellent book *Waste: Uncovering the Global Food Scandal*, 'the energy that goes into growing the 61,300 tonnes of perfectly good tomatoes that people throw into their household rubbish bins in the UK [could] grow enough wheat to relieve the hunger of 105 million people.'[17]

John's own journey into freeganism started in an unorthodox manner. He grew up in Folkestone, on England's south coast, and wanted to become a teacher. That all changed at college when, well: 'I had this massive dose of magic mushrooms,' John says, cheerfully. I can't help but laugh. John laughs, too. He is used to people's reactions to this story. After all, it's not often that someone you've just met will confess to their life-altering, definitely illegal, in-no-way-recommended drug trips, but John is an unusual case. Blame the mushrooms, or the years spent digging around in other people's bins, but he is less constrained by social taboos than anyone I've ever met.

The magic mushroom trip convinced John to, as he puts it, 'turn on, tune in, drop out'. He quit training as a teacher, and instead threw himself into the environmental movement.

John had always been conscious of his environmental impact in a noncommittal, everyday sort of way. But when he was tripping, he says, 'I realised everything is connected. You and I are made from the same stuff as those wooden doors, as those walls, as the cat down the alleyway, as the clouds, as the sun, as the universe. We're all made of the same stuff.' (I feel I should interject to point out that John is both entirely sober and entirely sincere as he says this.) He became a freegan, and he gave up eating meat. 'I would talk to a horse whilst I was under the influence, and I realised then that animals are sentient – something different, but not lesser,' he says, with a throwaway tone, as if it's every day you decide to give up meat after hallucinating a conversation with a horse.

These days John describes himself as a 'psychonaut', although his forays into psychedelics are much less frequent. 'I'm somebody that really enjoys experimenting with my own consciousness,' he says. Other aspects of John's life are equally unconventional. For one, he practices polyamory – that is, he has relationships with several partners at the same time – and that morning in York he spends several hours filling me in about his sex life in colourful detail (for another book, perhaps). Needless to say, John's lifestyle choices aren't for everyone. Still, few people have experienced the challenge of food waste as close up as he has, and I've come to witness what he's learned first-hand.

The shop's manager, Vicky, comes outside to greet us. A short, blonde woman in a puffer-jacket, Vicky points out some pallets of food that she has put out back especially for John: nectarines, oranges, cauliflowers. John picks up one of the latter, its leaves pale and slightly tired-looking, but the flesh bright white and enticing. 'I'll be having a cauliflower cheese made out of that,' he says.

There are many reasons that shops throw away food. Fresh produce might be damaged; tins get bashed and broken, by which point few shoppers will buy them. Shoppers in the Global North, in particular, are used to supermarkets projecting an image of bounteous plenty, shelves overflowing with nearly any kind of food you can imagine, no matter how exotic or out of season. Nobody likes empty shelves, and if something isn't in stock, then shoppers may decide to go somewhere else. The way that food is marketed and sold means it's often difficult to buy just the food we need. Instead, retailers and marketing departments spend billions every year working out ways to encourage us to buy in bulk – which makes food cheaper to buy, but ensures more goes to waste. Research has shown that heavy discounting and deals like 'Buy One, Get One Free' prompt shoppers to buy more than they will actually eat, thus generating a disproportionate amount of food waste.[18] What's more, much of the food we buy is portioned according to the needs of the average household – the fabled nuclear family of four* – rather than an individual, a problem in a society where more and more people live alone.

Food waste also creates problems with *other* kinds of waste. Head to the grocery aisle, and it can be hard to find a single edible thing that isn't wrapped in clingfilm and/or served inside a plastic tray. For many years, food companies have argued that plastic wrapping extends the life of fresh produce, and therefore reduces waste. The reality is more complex. Although researchers have shown that plastic packaging such as individually wrapped cucumbers can reduce food waste at retail,[19] a

---

* Don't get me started on 'portion sizes', which are little more than a food industry accounting trick, so that they can display the caloric values of an eighth of a pizza that they know you will eat all of.

recent eighteen-month study by the British sustainability charity WRAP found that in most cases, plastic-wrapped fruit and
veg made no difference to shelf life, and actually *increased* waste
overall, by forcing consumers to buy more than they need.
(The researchers found that by selling just five items – apples,
bananas, broccoli, cucumbers, and potatoes – loose, British
retailers could prevent 60,000 tonnes of food waste, and 8,800
tonnes of plastic, every year.)[20] Plastic, it turns out, appears to
be excellent for reducing food waste within supermarket and
food companies' supply chains – limiting their environmental
footprints, and bolstering their profit margins – but in aggregate, simply acts to move the moment that food is wasted onto
customers.

In recent years many retailers, and particularly the large
supermarket chains, have started to clean up their act when
it comes to wasting food. This is partly down to reputational
pressure, and partly due to legislation: the 2015 Paris climate
agreement, signed by 196 countries, set a global target of cutting food waste in half by 2030. France recently passed a law
requiring that all food waste be donated or fed to animals. In
Italy and Spain, restaurants are now required to package up
leftovers so they can be eaten at home.[21] The pressure seems,
at least on the surface, to be working: several big box stores,
including Walmart in the US and the UK's four largest supermarket chains, have set aggressive zero-waste targets and
pledged to send zero waste to landfill. Instead, unsold and
wasted food is increasingly being used for animal feed, sent for
anaerobic digestion – that is, turned into biogas, which is used
to generate electricity – or donated to food charities.

After a couple of hours of bin-diving and collecting donations,
John and I cycle over to Planet Food, one of York's food banks,

which is run out of a nearby Methodist church and community centre. Inside, it's set up for lunch. The wooden floor is dotted with dining tables dressed with yellow tablecloths; papier-mâché animal heads gaze down from an artist's studio upstairs. It is shockingly, heartbreakingly busy: young people, old people, disabled people, men in work boots and high-vis, a young mother feeding her daughter, who looks happily oblivious. Most of these people, John explains, are not homeless – though he helps to donate to the homeless, too, through the freegan charity Food Not Bombs – but everyday locals who just happen to be struggling.

Alongside the far edge of the room, a bank of trestle tables is piled with food donations, and minded by a crew of volunteers, almost exclusively elderly women. Planet Food operates on a 'pay what you can' principle. Most of the produce is donated by local supermarkets – Marks & Spencer peanut butter, Tesco popcorn, loaves of bread and salad from Waitrose. (Many supermarkets in the UK now have food bank collections, and encourage shoppers to buy extra food to donate – an easy way to look charitable while conveniently boosting sales.) There are spreads and cake made by volunteers, too. In the kitchen, another set of volunteers is cooking hot food: jackfruit burgers, potato wedges, soups, cheesecake for dessert. Steadily, people snake in the front door and load up a plastic bag with donations before sitting down to eat. We find a table and order. While we're waiting, a few people come over to say hello to John, sing his praises, or hear the latest on his sex life. The food, when it arrives, is nourishing and delicious. A sign on a blackboard says, 'Made with love and care.'

There are at least 2,000 food banks in the UK. In 2021, they served 2.5 million people. The number of people in food poverty in the country has soared in the last decade, exacerbated by cuts in government spending and a cost of living crisis.[22] FareShare, the country's largest food donation charity, served 131.9 million meals in 2020. The same challenges are felt elsewhere, too: in the US, Feeding America distributed 6.6 billion meals, 1.7 billion of which were due to retail donations.[23]

Supermarkets, John says, are not simply donating their food waste out of generosity. 'They used to have to pay for it to be taken away,' John says. 'Now what they do is the supermarkets put their wasted stuff aside, and they give it to a place like this' – he gestures around Planet Food – 'and the staff come and *collect it* from the supermarket. And they can pat themselves and the customers on the back, "you're doing a great thing, we're giving to charity, we're giving to the needy people!" Meanwhile, *thank you very much*, they write off the cost of the item at retail.' Food donations are at least in part a case of waste offshoring: for supermarkets, every kilo of bread or fresh produce donated to a homeless shelter or food bank is a kilo they don't have to pay to incinerate or dispose of in landfill. Many countries now incentivise retailers to donate excess food. In the US, for example, so-called 'Good Samaritan' laws protect organisations that donate food from legal liability, so that they cannot be sued if a homeless person gets sick from eating out-of-date sushi. Even then, the amount of food donated by retailers is still a thin slice of what actually goes to waste. In the UK, the biggest supermarket chains donate just under 10 per cent of their 'surplus' food.[24] The other 90 per cent is either pulped and sold for animal feed, or sent for aerobic digestion to generate energy – often for the supermarkets themselves. Both of these count as 'waste reduction', and count nicely towards companies hitting their zero-waste and carbon emissions targets, but are even more helpful for the supermarket's bottom line.

Looking around Planet Food at all these desperate people – most of them working, or in need of care – I once again feel conflicted about donations. Is it unequivocally a good thing to divert food that would otherwise be wasted, and to feed hungry people? Absolutely. But food donations can also help to mask wider problems, both at the societal level, and within our food system. FareShare estimates that its food banks save the British state £44 million per year[25] in costs that might otherwise fall on the NHS, justice and social care system. Depending on your own political beliefs, that's either a good thing (philanthropy!) or one that papers over serious failings within our social safety net. It feels good to feed people our leftovers, but I can't help but imagine, as I watch the queue for free food snake out of the door, how much better it would feel if we helped them get secure jobs that paid them enough to feed themselves. As for supermarkets, if they really want to help to eliminate food waste, they would be better focusing their efforts somewhere else: farms.

It's a typical early autumn day in Kent – tits gorging in the brambles, sun playing coy behind a bank of passing cloud – when I pull up at Boundary Farm. The scene could be snatched from a pastoral painting: white stucco farmhouse, apple orchard stretching into the distance. Scent of ripe fruit, low putter of tractors in the lanes. It's harvest time, and the farmer, Trevor Bradley, is out working the fields. But he is not who I'm here to meet. Instead, I've been invited by the food charity Feedback to help out with the harvest that happens *after* the harvest. I'm here to meet some gleaners.

Gleaning is an ancient practice. In the Old Testament of the Bible, God (via Moses) instructs farmers not to harvest the edges of their field, but instead to 'leave them for the poor and the foreigner'. The Quran, similarly, dictates that a tenth of the harvest should be donated to the poor. This second harvest, gleaning, was a common practice in farming communities all over the world for hundreds if not thousands of years. In medieval Europe, landowners would leave the gleanings of the harvest for the farm workers and the village poor, ringing a bell towards the end of the day to signify that the gleaners could take their fill. In this way, the surplus and wasted food would be eaten – and the rich could cleanse their conscience. In the last hundred years, however, the practice of gleaning has fallen out of popularity, or at least out of sight. Most of us, thankfully, no longer have to till the fields at harvest time. In the Global North, farming is an industrial behemoth, the harvest mechanised or otherwise contracted out to low-paid and itinerant migrant workers. But in recent years, charities and environmental groups have begun to resuscitate the practice of gleaning, collaborating with generous farmers to collect unused and surplus food and distribute it to food banks.

Charmaine Jacobs, Feedback's local coordinator, meets me at my car. A bubbly young community college teacher with cornrows and bright red glasses, Charmaine has been gleaning with Feedback since 2015. She first heard about gleaning, and the problem of food waste, while studying for a business management degree at the nearby community college. 'I thought, "That sounds like fun, I love being outside," and then I got the bug,' she says. I can feel Charmaine's enthusiasm as we cross the farmyard, a cluster of dusty green corrugated outbuildings separated by a stretch of hardstanding. The group is already getting stuck into several large wooden crates of potatoes, some stacked above head height; there must be several tonnes here, although nobody is keeping count. Boundary Farm is a sizeable operation: 650 acres of winter wheat, potatoes, and year-round brassicas, including cauliflower, cabbages (white, red, Savoy), and Brussels sprouts around Christmas time. Trevor, the land-owner, inherited the farm from his father, and has turned it into a reliable and well-loved producer, selling veg boxes to farmers' markets and retailers for miles around.

'These are chipping potatoes, used for making chips,' Chris Turnbull, one of the most senior members and de facto leader of the group, tells me. Chris is a thin stick of a man, with a brush of silver-white hair and lusciously thick eyebrows, his T-shirt and cargo shorts dirty and well-worn. A former medical engineer, Chris spent his career designing tracheotomy tubes and catheters – 'A tube for every orifice, as they say,' he jokes. He and his wife Sue (straw-blonde hair, lemon-yellow blouse) fell into gleaning after retirement because they wanted to do something to fight climate change, and this felt like an imme-diate, practical way to make even a small difference – as well as being a nice way to spend time outdoors.

Chris offers to show me the sorting process, so we stroll over to a nearby farm building. Inside one of the lofty corrugated barns is a long red machine which looks like the mutant off-spring of a conveyor belt and a colander. The belt and the floor are littered with enough discarded potatoes to fill a grocery aisle. This is an automated sorter. Every day, Chris explains, the farm workers haul truckloads of potatoes from the fields and dump them onto the sorting machine to be graded: sorted by size and shape, and the rejects removed. The chosen will be sent for further processing. The rejects . . . well, that's why I'm here. 'They don't want ones that are weird, wavy shapes, because when they go through the chipper you get odd sorts of cut-offs,' Chris says. He picks a potato off the conveyor. The skin is pale and waxy, and fresh loam crumbles off under his fingertips. A little small, perhaps. But otherwise beautiful, fresh, delicious-looking.

'See?' Chris says. 'They're absolutely perfect.'

Back at the crates, I don my gardening gloves and get stuck in. Some of the potatoes have been left in the sun, and so are greening; others show signs of moulder, a consequence of being left lying out too long in the rain. 'None that have got really deep cuts, or have been sliced by the farm tools, like that,' Sue says, brandishing one that looks like Mr Potato Head had a date with the guillotine. 'If it's only got a small bit of green on it, they're OK.'

I lean into the harvest, inspecting each spud, mildly intoxicated by my newfound power as judge and executioner. (The good shall be saved, the unrepentant and mouldy cast out.) There are heart-shaped spuds, lung-shaped spuds, potatoes pocked with holes. Some are the size of a coconut, others the size of golf balls. Plenty of them are indeed green or damaged, but perhaps a third of them are indistinguishable from perfect, and another fraction evidently edible – the kind that, if you found them in your fridge, you'd just chop off any bruise marks and toss them in a roasting dish. It's hot, sweaty work, hard on the back. Luckily, Sue is a retired physiotherapist, and helpfully corrects my posture. After an hour or so, the act becomes routine, and we all get faster. In fact, whereas at the beginning finding each salvageable potato gave me a quiet thrill, after half an hour I begin to realise that it is actually harder to find potatoes here that you couldn't eat. These weren't waste; they were being wasted. I say this to Chris, who by now has clambered into a crate, waist deep. He nods as if to say: *I know.*

'Last year we did 35 tonnes,' he says.

Despite being an open secret, good data on how much food is wasted at harvest is difficult to come by. You can understand why: if you're a farmer, admitting to an oversupply of a particular fruit or vegetable is only going to tank your price at market.

In the UK, Feedback estimates that between a quarter and a third of food waste, up to 5 million tonnes per year, may occur on farms. The problem is likely greater than we had previously estimated. A study by scientists at Santa Clara University recently found that a third of all crops in California are wasted at harvest level.[26] It's not just arable farming, either. Fifty-five million tonnes of milk never make it to the shelf.[27] One in three fish caught is either thrown back or left to rot on the decks of fishing boats, according to the UN Food and Agriculture Organization.[28] The World Wildlife Fund estimates that 1.2 billion tonnes of food – or 15.3 per cent of all food produced worldwide – is wasted before it even makes it to the store.[29]

Farmers hate to call leftover food 'waste'. Most would rather call it 'surplus', and it happens for all sorts of complex reasons, almost all of which boil down to one simple one: money. Farming, even on modern industrial megafarms, is still an unpredictable business. It relies on making a bet – selecting which crops to grow, preparing the fields, tending them laboriously through the year – on uncertain returns. One bad growing season (too dry, too wet, too cold, too hot) can upend months of effort. A glut can be almost as bad: if the market for a particular crop becomes saturated, prices plummet. Should prices fall low enough, a farmer might decide that the crop is no longer worth the cost of labour required to bring it in, and so entire harvests are left rotting in the fields, or ploughed back into the soil, only for the farmer to then start all over again. In 2017, for example, a hot spell in the UK led to an oversupply of cauliflowers, which farmers were leaving to rot in the fields until supermarkets, under pressure from NGOs, agreed to increase their stocks to save them from spoilage.[30] The previous year, US dairy farms poured away an estimated 195 million litres of surplus milk due to oversupply.[31] 'I've got friends in

Cornwall this year that bypassed 600 acres of crop because the weather brought the crop on too quickly and there wasn't the demand,' Trevor, the farmer, says. 'The problem is big growers can't afford not to have the product on the shelf. They tend to grow 20 per cent more than they need, so therefore at times you've got massive extra crop.'

The problem is compounded in poorer countries, and in warmer climates. Even before it is harvested, food begins to decay, a process only expedited by heat and moisture. Getting food from farms and onto our plates is a race against putrefaction. To combat spoilage, we have spent a century building a complex supply chain – washing, drying and packaging food into refrigerated trucks and shipping containers – which helps to extend a product's freshness on its journey from farm to plate. These supply chain technologies have successfully created, at least within the brightly lit aisles of Western supermarkets, the illusion of a world without seasons, one where a punnet of blueberries or a packet of avocados can be on our plates within days of being picked, oceans away. In many poor countries, however, many of these technologies are not widely available, and so wastage on farms and in open-air markets is even higher. As many as half of all avocados grown in the Global South become waste before they reach market. The rate for citrus fruit is likely to be higher.[32] In fact, according to the FAO, in the Global South, 40 per cent of losses happen before food reaches the market.

Once a crop is picked, it must be graded. Supermarkets and food processors judge a crop's quality (and therefore its price) on everything from its size, to colour, to sweetness. Grading standards don't always come from manufacturers. The European Union, famously, dictates marketing standards on a number of fruits and vegetables,[33] which led to the notorious myth that it banned the sale of straight bananas.[34] (The truth is more complicated.)[35] But the real burden of cosmetic standards comes from supermarkets, and from big food brands, which control most of the food market. According to research by Feedback, farmers lose up to 40 per cent of their crop due to not meeting supermarket cosmetic standards.[36] A study of tomato farmers in Australia found that up to 86.7 per cent of undamaged, edible tomatoes were rejected as outgrades.[37] 'They want everything perfect,' Trevor says. 'I know some parsnip and carrot boys up in Norfolk. The supermarket wants exactly the same shape parsnip in their packs. You grow some parsnips in the garden and see how many you get that are the same. It's not an easy job.'

'So when you talk about potato packing, yes they might pay a bit more money,' Trevor says. 'But every single potato has to be perfect. And to get a perfect crop in the changing climates we have is very difficult.'

For years, British farmers have also complained of exploitative treatment by supermarkets, who will often amend or cancel contracts on short notice due to inaccurate sales forecasting (products selling less than expected) or sudden changes in demand (for instance, when a summer wet spell depresses the demand for ice cream and barbecue food). In the UK, such abuses led to the creation of a specialist body, the Groceries Code Adjudicator, which seeks to prevent exploitative treatment by retailers. Despite this, one in five farmers still reports having contracts amended or cancelled on short notice – which, inevitably, leads to greater food waste.[38] And elsewhere, the problem persists.

Trevor no longer works with supermarkets. 'I think they've been the demise of a lot of businesses,' he says. Over the years, he's seen the farms around him having to sell up, no longer able to make a living. A few of the local farmers formed a co-operative in the village to try to negotiate better deals, but the price pressures, a shortage of affordable labour, along with the growing unpredictability of the weather due to climate change, have made working the land untenable for many. 'We're one of the last people down this way doing it in any quantity.'

Another group of gleaners pulls up in a tired-looking minivan, returning from a nearby orchard. Charmaine cracks the boot, and I'm hit by the acid-sweet aroma of hundreds of apples, hand-picked and placed into net sacks. Charmaine offers me one. They're Worcester Pearmains, an old English variety with roots that date back to the nineteenth century; tart-sweet, their flaky skin a bright lime green that gives way to a deep crimson flush. Hungry from the work, I take a deep bite, and an unexpected burst of juice dribbles down my chin. It's delicious.

We decide to break for lunch and sit in the grassy orchard between rows of fruit trees. Home-made sandwiches are unwrapped from their foil, cold drinks poured. The sun throws a dappled shade through the trees. The odd bee drones by, perhaps curious at the sight of us. It's all strangely idyllic. Some of the gleaners are swapping jam recipes. George and Chris are talking about the hop harvest; a few of them have formed a collective to make beer. It's hard not to notice that of the ten or so gleaners here, Charmaine and I are the youngest by at least a couple of decades. 'I think pretty much everybody's retired,' Chris says. Most are like him and Sue – people looking for a way to help out, but also to fill their days. This is the core group, the hardcore, although the list of volunteers stretches into the hundreds, and will rotate in and out, depending on the day and the season.

I think briefly of the actual pickers who bring in the real

harvest, generally migrant workers brought in by agencies from abroad, and who work under relentless conditions for low pay. Nobody here is getting paid at all, a fact that I am increasingly uncomfortable with. In one light, it's good old-fashioned philanthropy, done in the spirit of giving. In another, it's waste disposal built on unpaid labour. That isn't to say that gleaning feels exploitative – far from it. It's clear that Trevor hates to see anything grown on his farm going to waste. He is generous and welcoming to the gleaners, which isn't true of all farmers. ('Some of them won't open up their farms to anyone,' Sue says.)

Chris and Sue have picked almost everything that grows in this part of the country: summer fruit, stone fruit, tubers, brassicas, carrots, squashes. True gleaning takes commitment. There is food waste to pick and hungry people to feed all year-round, and in all weathers.

I ask what makes for the hardest harvest.

'Potatoes in December, when they're wet and slimy,' Sue says.

'The worst,' Charmaine agrees, although she for one prefers potato gleaning to hot days picking stone fruit: peaches, plums, cherries. 'You get wasps.'

'I'd rather deal with a wasp than that soggy potato smell,' Sue grimaces.

When it comes to waste, Charmaine says, potatoes aren't the worst. That unfortunate honour would go to pumpkins. Around 80 per cent of the 10 million pumpkins grown in the UK are used as Halloween decorations and then thrown away uneaten, creating an estimated 18,000 tonnes of food waste.[39] (The US, with its strangely deep obsession with Halloween, throws away 900,000 tonnes.) Any unsold crop that the gleaners don't take is discarded or dug back into the soil. 'You look at the field,' Charmaine says, 'and you just think: *how?*'

* * *

Of course, to really make a dent in our food waste problem, we need to change how we eat at home. According to the food charity WRAP, 70 per cent of all post-farm food waste in the UK arises not in retail, but in our kitchens.[40] Some of that, undoubtedly, is down to the actions of food manufacturers and supermarkets: portion sizes, Best Before dates,[41] marketing deals enticing us to buy more than we really need. But it's largely down to our own relationship with food. As modern agriculture and supply chains have made food cheap, diverse, and plentiful, it can also sometimes feel that we have forgotten to value the food we do eat, or understand the environmental or human costs that have gone into its production – from meat farming to out-of-season blueberries delivered by refrigerated air-freight fresh from Peru.

Chris, Sue and the group have seen this change for themselves. 'Even when we give them away, we've had clients who say, "Oh, the potatoes are quite *muddy*,"' Sue says, irritably. 'I say, "Yes, they came out of a field. You *wash* them."'

There's a murmur of agreement among the group. 'We've now got to break the cycle of wanting clean, ready-prepared, washed vegetables, which aren't created that way,' Chris says.

'Nobody's life is so busy you can't cut up and wash a cauliflower,' Sue says. 'We've lost the plot.'

In their work studying landfills in Arizona, William Rathje and the garbologists at the University of Tucson found that families wasted less food when their diet was consistent – that is, they ate a smaller range of foods more frequently. One-off or unusual purchases (say, cooking a new recipe, or hosting a dinner party) led to more wastage. Rathje called this 'The First Principle of Food Waste', and it is perhaps more simply summarised as: *buy what you always eat*. Poorer families, they found, bought smaller portions, and thus consumed more packaging, but wasted less food. The most food, paradoxically, was wasted during periods of crisis. The reason, Rathje surmised, was panic buying – people bought more than they needed, which inevitably ended up in the trash.[42]

Many of the best ways to reduce food waste at home are also the simplest: buy smaller portions. Buy locally, and direct from the farm, if you can afford to. Buy in cans or frozen where possible. (The food is often fresher that way, anyway.) Refrigerate fresh fruit and vegetables, except potatoes, which are best kept in the dark. Little tricks, like keeping a lettuce in water, as you would cut flowers, instead of loose.[43] Using food scraps to make soups, or smoothies, or even – as John Cossham would tell you – fruit leather. I'm not advocating that everyone return to the lifestyle of a 1920s housewife, with pots of bone broth constantly on the boil. New technologies can help, too: food-sharing apps like Olio and Too Good To Go, which let shops and households in the UK advertise spare food to share among the local community, are undoubtedly good things. Similar movements are spreading globally.

Spending the day toiling in the fields with Chris, Sue, Charmaine and the others, I'm filled with a renewed affinity and appreciation for food: the crumble of the soil in my fingernails, the flake of apple skin. After lunch, we jump in the cars and head down a series of narrow and picturesque country lanes to another part of the farm, into the brassica fields. A spell of hot weather had brought the cabbage crop on earlier than expected, while the hired pickers were busy harvesting other, more marketable produce, and so there are still swathes of the crop left unpicked and going over – which means we can have our fill. I look out at what must have been several thousand cabbages, layered green heads bursting from the earth like wrinkled flowers. Chris hands out gloves and what appear to be bread knives, serrated and several inches long, before demonstrating the proper technique. He crouches, gripping a cabbage head firmly with one hand, and saws through the stem at the base with the other, in a move grimly evocative of a scalping, or decapitation. Where the cabbages' outer leaves had crisped and withered in the sun, they too were to be removed, and cast to the ground, where they'd be ploughed back under to nourish next season's crop.

We set to work. Nearly every inch of the soil is covered with leaves, which crunch crisply under foot. Bending low — incorrectly, judging by the blooming pain in my lower back — I quickly get into a rhythm, sawing and yanking. The cabbages are heavy, loaded with moisture, and almost all of them, aside from a few outer layers, perfectly edible — saleable, even. Soon we are filling crates by the dozen. We must make a peculiar sight, this line of knife-wielding pensioners advancing ominously through the field like extras from *The Wicker Man*.

Late in the afternoon, with the gleaners' cars weighed down with fresh fruit and vegetables to be delivered to needy families, we say our goodbyes. In lieu of payment, Charmaine and Sue send me off with a small sample of the spoils: a cabbage, four apples, a few massive potatoes. That night, back at home, I bake the potatoes and serve them for dinner, with a little salt and lashings of butter. For dessert, I cut into one of the apples, which still carry that tangy smell of the orchard. Even my 4-year-old — who only eats most fruit under extreme duress — agrees that they are delicious.

After dinner, I pile the scraps into the food bin, which is full to bursting.

8

BREAKDOWN

*'The trash and litter of nature disappears into the ground with
the passing of each year . . . man's litter has more permanence'*
—JOHN STEINBECK, *The Pastures of Heaven*

In York, after lunch at Planet Food and a quick stop to dive
another bin ('Fucking hell, lead!' John says, loading metal onto
the trailer excitedly), we cycle over to John's house on the
other side of the city. It's a ramshackle place, old and charming. John leads me through a side gate, past a pile of scavenged
bric-a-brac, and out into his garden. Grass and weeds climb
waist high. The vegetable patch lies lush with spring bounty:
perennial spinach, sorrel, asparagus. A redcurrant tree sags
under the weight of fruit. I feel like a character in a children's
novel who has wandered through a door into a hidden paradise.

'I'm afraid it's chaotic and untidy,' John says.

It's lovely, I say earnestly, waving away a cloud of insects.

'I love the chaos and abundance of nature, so other than
doing a little bit of management, I like leaving it. Other people
would say this is an untidy place, and yes, it's untidy,' he says.
'But it's productive.'

We reach a clearing at the bottom of the garden. The food
waste that John cannot redistribute or eat himself he brings

here, to make into compost. All around the fence line are compost bins: black bins, green bins, some hacked together out of wood pallets, looming like a mob. Most of the compost John produces is used around the city in community gardening projects, the rest he sells at a local farmers' market. 'Every ten kilos of food waste makes just less than one kilo of finished compost,' John explains. In a typical week, he might lug back 150kg of food waste – a tiny fraction of what a commercial composting business might do, but plenty to keep one man busy. John has been making compost now for more than a decade. As an accredited 'Master Composter', he estimates he has trained thousands of people over the years in the delicate art of decay.[1] (You don't get the name 'Compost John' for nothing.) As someone with a newfound obsession with food waste, I'm hoping he'll give me some pointers.

What do we do with all the food waste that we cannot avoid? An honest answer might be 'not enough'. Globally, organic waste makes up 44 per cent of what we throw away, and yet we compost only 5.5 per cent, according to estimates by the World Bank.[2] (Anaerobic digestion to make biogas – on which more in a minute – doesn't even crack one per cent.) Although in the UK the majority of our food waste is either digested or composted, nearly 30 per cent remains unaccounted for, which means that it is likely buried or burned.[3] That rate is even higher elsewhere: Australia sends nearly half its organic waste to landfill. The US (them again) buried or burned 81 per cent of the food it threw out in 2018.[4] I find those figures particularly dumbfounding, because organic waste is the only thing we throw away that comes with its own free, inbuilt disposal method: rot.

Nature knows no such thing as waste. When our cells produce waste by-products, microscopic organelles called lysosomes remove and digest them, providing the body with energy to grow new cells. As vertebrates, our bodies harvest excess calcium to build our bones. (As Steven Johnson has written, 'Your ability to walk upright is due to evolution's knack for recycling its toxic waste.')[5] When a tree takes root and sheds its leaves, it feeds the soil it grows in; when the tree bears fruit, that fruit is eaten and eventually excreted by animals, returning those nutrients to the soil. Just as one man's trash is another man's treasure, so one organism's crap is invariably another one's dinner. This basic truth, the circle of life, is something we learn in primary school (or the first time we watch *The Lion King*). In ecology, such systems are referred to as closed loops,* and they rely on a simple biological tenet: more or less everything degrades.

Humankind has long been disrupting this natural cycle. When we grow food, it is no longer consumed and returned to the earth it was grown upon, but often shipped across the globe to be eaten in some foreign city, whereby the waste (in the form of both food and human waste) is not returned to the soil, but ejected via sewers into the rivers and seas. The lost nutrients must therefore be replaced with synthetic fertilisers, contributing to our highly productive[6] – and extremely polluting – food system. In *Capital*, Karl Marx describes this dislocation as a metabolic rift, which 'prevents the return to the soil of its constituent elements consumed by man'.[7] To Marx, this simple cycle – feeding the soil that feeds us – was 'prescribed by the natural laws of life itself'.

Soil – 'the thin cushion between rock and air on which human life depends', as the environmental journalist George Monbiot has so elegantly written[8] – is an ecosystem founded on waste. Soil *is* decomposition: a single gram of it can contain

---

* In contrast to the open loops of our current predicament, in which materials are regularly lost to landfills, or the atmosphere.

billions of bacteria and a kilometre of filamentous fungi, all of them breaking down organic matter and excreting chemicals. (These emissions – bacteria breaking wind – give mulchy soil its unique smell.) These microscopic populations create the soil's internal structure, depositing carbon via cement-like polymers, and enriching it for other plant life.[9] When farmers till fields, it destroys these internal structures, and species diversity plummets. Compacted and overtilled soil absorbs less water, increasing run-off, and increasing topsoil erosion. As soil quality has declined, the tendency has been for farmers to add more and more synthetic fertilisers to try to improve yields. But the majority of these fertilisers – almost 60 per cent of nitrogen and up to 80 per cent of phosphorus sprayed – are wasted.[10] Instead, the chemicals are carried via runoff into rivers and streams, contributing to hypoxic underwater 'dead zones', some of which, such as that found in the Gulf of Mexico, are now thought to span thousands of square miles.[11] According to the UN, a third of the planet's soil is severely degraded, and is being lost to erosion at a rate of 24 billion tonnes a year.[12]

The world is only now reawakening up to the role that waste plays in our soil. The rise of no-till and 'regenerative' farming has meant that the wider scientific and agricultural community is re-evaluating the role that decomposition might play in soil health. And that means compost.

Composting is nearly as old as farming itself. There is evidence of Scottish farmers composting as early as 12,000 years ago; Cuneiform tablets discovered in Mesopotamia record the practice in 2,300BC, during the reign of King Sargon of Akkad.[13] According to the journalist Heather Rogers in *Gone Tomorrow*, large-scale industrial composting was developed by an Italian doctor named Giuseppe Beccarin in 1914, and was initially known as 'fermentation'.[14] Before that, there was no real commercial need for such a system. The processing of organic waste was no different from the processing of human waste – everything fed the soil.

I had always thought of composting as a niche practice, a nice hobby for hippies in hemp T-shirts to conduct at their allotments. Then a few years ago, I read a *New York Times Magazine* article about two California 'carbon farmers', who were using compost to help increase soil's ability to sequester carbon.[15] The evidence is growing: in 2019, researchers from the University of California, Davis published the results of a nineteen-year study which found that adding compost to soil alongside cover crops increased the soil's carbon content by 12.6 per cent.[16] Rather than letting our food waste simply emit carbon into the atmosphere, composting it might help us return it to the earth.

The key to good compost, John explains, is rotation. We stop at a bin to cram in some blackening salad leaves. 'This is the decomposition stage,' John says. Inside are decomposing rinds and peels crawling with insects. 'It'll have six months to a year in here, and then I'll move it all over to this one.' He opens its neighbour, in which the waste has broken down into a black and lumpy mulch. 'This is the maturation phase.'

In John's system, food waste is rotated through at least two composting cycles. The natural breakdown that happens inside a compost pile relies on an entire ecosystem of hungry organisms – bacteria, fungi, archaea, ants, flies, worms, beetles – to break down and consume the food waste, excreting nutrient-rich humus. This technique, vermicomposting, is an aerobic process; unlike the anaerobic breakdown inside a landfill, it requires oxygen, but produces fewer malodorous gases, and far less methane.

John pops open the lid of a wormery. The scent is abrupt and sour, and I have the sudden urge to retch. 'Lovely wormies!' John says, sticking an ungloved hand in to rootle around. John farms worms both to process his own compost, and to sell to others. 'Those are tiger worms – can you see, they're slightly stripy? Those white things are pot worms.' He gestures to another cluster of white writhing *enchytraeidae,* then stops short to admire a fly. 'Oh, lovely little darling! This is one of my favourite insects, a dipteran fly – a two-winged fly. Its larva is called a rat-tailed maggot. They like to live in dirty water and in manure, and make *terrific* compost.' Inside the wormeries themselves (John has several, each tiered like wedding cakes) the worms digest the biological waste and excrete it into a fine, wet mulch. 'This will dry out and become really rich compost,' he says, pulling out a tray and stirring the mixture, as if reducing a ragu.

Over at another pile, John shows me his secret ingredient. 'This is a biochar walnut shell,' he says, holding up what looks like a small lump of charcoal. Biochar is organic waste that has been pyrolised – burned in the absence of oxygen. Among farmers, biochar is sometimes considered something of a miracle material. 'It's carbon sequestration,' John says.[17] 'Photosynthesis takes carbon dioxide out of the atmosphere and makes it into carbon, long chain starch molecules. When you compost something, about 50 per cent of it goes back into the atmosphere as respired carbon. But you cannot degrade charcoal – it's stable in the soil. So you can use photosynthesis to drag CO_2 into plants, biochar those plants, put them in with your compost and compost becomes much richer and better for plants to grow.' John makes the biochar in a retort (a kind of oven). It's low-yield but, he says, makes the compost both higher quality and better for the planet.

Our final stop in this micro-tour is the one that I had been anticipating with what might, to an ordinary person, seem like a strange level of excitement.

'That's my compost toilet,' John says, nodding towards a wooden structure. 'Would you like to see it?'

'Sure!' I say. Even then, I'm not quite ready for John's candour.

'That was deposited yesterday, and that's Helen's,' John says, indicating his wife's apparently most recent bowel movement. My own disgust is misplaced: there's surprisingly no odour, and little in the way of mess. After each occasion, John and his family simply cover their shit – known as 'humanure' – with woodchip, leaving it to decompose. 'This is finished, humanure compost,' John says, leading me over to an open bin of rich, dark humus. He pinches a handful of what was until recently faeces, and rubs it between his fingers. I must confess, it looks surprisingly lovely: rich and crumbly, indistinguishable from good compost. 'That will take eighteen months to fill up,' he says. 'I use it under trees, hedges, that kind of thing.'

'Aren't you worried about pathogens?' I ask.

'China has been composting their animal mortalities and humanure for four thousand years,' John says. 'Pathogen removal is a combination of temperature and time. So if you get it up to 66 degrees Celsius, and hold it there for one hour, that will eliminate all the pathogens. If you only get up to 55, you might have to hold it there for twenty days. This is around 20 degrees, but that's going to sit there for the next three years. That is now being worked on by worms, mites, bacteria – a whole ecosystem going on in there. And there is no way that pathogens will survive.' Composting toilets aren't for everyone. They have, however, been shown to be a safe and effective means of human waste disposal, particularly in the Global South and in agricultural communities, where there is an immediate use for their compost by-products. In rural areas, composting toilets are a compelling alternative to septic tank systems, providing sanitation and fertiliser all in one.

To John, compost is more than just a way to process food waste. It's an ethos. Not one that was delivered to him by a talking horse, but by a sense of the interconnectedness of nature. 'People don't understand that we are part of nature, and that this is a natural cycle. The sooner we can get that we need to fit in with nature and stop dominating it, but manage it in a way that is sympathetic to the natural processes, the better.' Too often, he says, people see compost as just soil – something to be bought in bags at the garden centre. People associate the word with being good for the planet, but it is not always the case. In the UK, more than a third of all compost sold is peat compost, dug from among the richest carbon sinks in the world. Globally, peatlands hold twice as much carbon as all the world's forests combined.[18] In digging up our peatlands to grow Instagram-friendly monsteras and fiddle-leaf figs, we are hastening a climate catastrophe.

I find John's approach to living – waste as little as possible, and compost the rest – at first disgusting, then disorienting, and eventually powerfully persuasive. John has built his life around eliminating waste, but also on kindness: the little waste that he cannot share with the needy, becomes compost that will return carbon and nutrients to the earth. He's trying to heal the metabolic rift, one small action at a time.

Eventually it's time to leave. As I depart, John is on his knees, carefully picking teabag labels and plastic shards from his soil, before it's ready to be bagged. 'Look, hedgehog poo!' he says, beaming. 'Honestly – how can you not love compost?'

* * *

Energised after my trip to York, I decide to give composting a shot at home. As it happens, our house already has two large and plastic compost bins, inherited from the previous owners. When I first open them, they make for pitiful viewing. Dried out and dead-looking, with the telltale aroma of mould, they are more dust bowl than compost bin; the result of years of neglect. So, with John's advice in my ears and scent still in my nostrils, Hannah and I eagerly set out to resurrect them. In go grass cuttings, rose prunings, and an embarrassingly large amount of food waste: stale bread and potato peelings, apple cores, chunks of banana gloop scraped off high chairs. It turns out that children under five won't eat their broccoli even if you diligently explain the relationship between food waste and carbon emissions, and so we still have a steady stream of uneaten vegetables and gnawed crusts.

Every few weeks, we stick a garden fork in and turn the pile over, aerating it. I find myself eagerly Googling articles about Bokashi, a Japanese method of composting in which *Lactobacilli* bacteria are added to help break down the food waste. (Bokashi is to compost what sourdough is to baking; a once niche skill now undergoing a hipster boom largely thanks to Instagram.) After a few weeks, our compost is alive and – if not quite thriving – certainly decomposing. The humus turned darker and more soil-like. And as if from nowhere, a whole menagerie has moved in: ants and mites, lice, fly larvae, and fat, healthy worms. Watching them wriggle and pulse gives me a disproportionate sense of joy, to see such life amid decay. Surely, I can't help but think, this is how waste is supposed to be.

We do, however, quickly run into the problem of what to do with our new rotten crop. After spreading some of the crumbly new compost on our flower beds, we're quickly running out of ideas. (We have only so many plants, and Hannah will testify that I am useless in the garden.) After a couple of months, although the compost heaps are still healthy and well fed, I find our early zeal is flagging slightly. There's only so much compost a single family can use. Besides: our local council already has regular food waste and garden waste collections. I want to see what happens to all that food waste at scale, done properly. So after a phone call and email to the council I find out where my organic waste is headed: Cumberlows farm, in the nearby village of Rushden.

The owner of Cumberlows is James Hodges, a rugged-looking 52-year-old with wavy black hair, who strides out of a Portakabin to meet me in Prada sunglasses and shorts, despite the fact it's December. It's Christmas Eve, in fact – the only day that James could fit me in to his busy schedule. 'We've got a lot coming in at the moment,' James says. Compost is a very seasonal business. In the run up to Christmas, Cumberlows will see a big spike in food waste (it's the parties). Afterwards, there will be a glut of Christmas trees in varying states of disassembly. 'It used to be after Christmas, all we'd get is cardboard, Tetrapak, plastic, it was awful.' He grimaces. Now, thanks to local campaigning and better recycling collection, the contamination rate is lower, although stuff does still get through. 'Spring is our big time,' James says. 'Whenever people are cutting their grass we are busy. Our biggest risk is drought because nothing grows, so we lose money.'

James used to be a loss adjuster, but gave that up to help his brother run the family farm after their father died. 'It's not a big farm, under four hundred acres,' he says. They grow row crops: wheat, barley, oilseed rape. But James himself is not really interested in farming. 'Sitting and ploughing for eighteen hours gets really boring. Whereas here, you're doing something different all the time.' We wander over the farmyard. Ordinarily, there would be garbage trucks queuing into the weighbridge to unload, but today most people are off work, finishing up their last-minute Christmas shopping. James's crew are still working, though. Over inside one of the tall, open sheds, a man in a loader is shifting heaps of organic waste. Cumberlows processes compost waste for North and East Hertfordshire councils, as well as commercial customers. In total, the farm processes around 40,000 tonnes of organic waste per year. That makes it a small player in the market when compared to specialist facilities run by the big waste companies (Veolia, Biffa), but enough to sustain itself.

Organic waste is usually separated into green waste (that is, garden and agricultural waste) and food waste. The two are collected separately, and piled at opposite ends of the shed: at one end, a voluminous heap of what appears to be largely hedge trimmings, at the other a smaller but far more noxious pile of food waste mouldering in compostable bags. I recognise the hit of ammonia and volatile compounds instantly. A track loader is mixing the piles together, before passing it through a shredder. Watching the atomized muck firing out of the back end, I suddenly understand the smell.

We stroll next door, to an adjacent series of buildings: a grid of what are essentially large airtight boxes, big enough to park a couple of minibuses inside. These are what James calls 'clamps', or composting vessels. The chambers are sealed from the elements. Inside, an aeration channel stirs through oxygen, ensuring faster decomposition. 'Sterilisation means four consecutive days at 65 degrees,' James explains. 'It's higher because it's got food waste in it.' After an outbreak of foot and mouth disease in 2001, the UK passed legislation strictly limiting the use of animal waste in compost. 'You can no longer compost food outside. If you want to compost any animal by-products, it has to go through something like an in-vessel composting, or an AD [anaerobic digester],' James says.

After sterilisation, the compost spends a week at 45 degrees in another clamp. Then it's taken outside and placed on the windrows: great long piles, at least nine feet tall, like earthen levees. I can see steam rising from the waste in the pale December air. 'They can be generating heat for up to a year,' James says. I breathe in the scent, expecting something powerful, but it's surprisingly faint – loamy, like petrichor. 'It shouldn't smell if it's done properly,' John says. 'It still *smells*. But if you have a really bad acrid smell like ammonia, that's bad compost. It's gone anaerobic.'

By this stage in the process, the centre of the compost is teeming with microscopic life, which rapidly breaks down the food waste – proteins, fats, fibres, starches – into carbon-rich mulch. 'Between six weeks and twelve weeks is standard to make compost. We work on a seven-week process from start to finish,' John says. Every few weeks the piles need turning with a telehandler (like an oversized forklift). We wander between the rows, which rise on either side of us like trench walls. I can hear a robin calling off in the hedges, and the suck-slap of wet mud underfoot. The windrows, I notice, are marked with dates. 'It's for traceability,' James says. 'We know exactly where all this waste has come from. The idea is that we can follow it all the way back through the process.' If there were to be a rare disease or toxic chemicals detected within the compost, he explains, it narrows down the search. For every 5,000 tonnes of waste that passes through, samples must be taken and tested for pathogens and chemical contaminants. 'You used to get shred and spread cowboys,' James says. 'Now everything has to conform to BSI PAS 100' – a standard set by the British

Compost Association and the British Standards Institution that regulates compost quality. He gives me a weary look. 'There's an awful lot of paperwork.'

Compost is not an easy business. 'It's more profitable than farming,' James says – though these days, that doesn't count for much. While some composters are focused on selling product, Cumberlows is fundamentally a waste disposal business: though it does sell its compost to the general public, the majority is given away to local farmers for free. The business makes money on its disposal contracts with local councils. 'In the past, I literally couldn't give it away,' James says. Farmers want fertiliser; compost is not considered fertiliser, but a 'soil-improver'. 'It's the organic matter that's important. They're slow releasing, whereas a bag of fertiliser, it's going to do exactly what it says on the label, and quickly.' Lately, however, farmers have started to show a renewed interest in compost, due to the growing evidence of its role in improving long-term soil health. 'Now because of carbon offsetting, it's really helpful for the farmers to put the compost into their field. And they'll probably be able to claim a grant for that soon,' he says. And if anything is going to convince farmers, it's money.

In England, the challenge for composters is made harder by the haphazard means by which waste disposal is governed – waste policy is set not by central government but by 333 different local authorities. 'They all have their own agendas, their own budgets,' he says, irritably. Currently only 51 per cent of these collect food waste separately from the rest of the garbage.[19] Whenever budgets are tight, food and green waste is often the first to go. (This phenomenon is universal. New York, for example, has a long history of cutting and reintroducing its compost collection. At the time of writing, it is paused. Again.) This is partly political: while everyone throws food away, lawn and garden waste is mostly generated by the rich. James calls it 'the posh waste'. In the UK, many councils – including mine – now charge a separate fee for collecting garden waste, which is both a cost-saving measure and pitched as a progressive tax, as the wealthy are more likely to have larger gardens and the time to prune them.

It doesn't have to be this way. In South Korea, for example, compost is a fact of life. Not only is food waste collection universal there, compost bins stand on street corners alongside the regular trash cans and recycling. Its world-leading rot infrastructure is the result of a mandatory food-waste recycling policy that the South Korean government launched in 2013 in order to tackle its then-famously high levels of food waste. The country now recycles 95 per cent of it – up from just 2 per cent in 1995 – via a combination of biogas and compost, the result of which is used to feed a blossoming network of rooftop gardens and urban farms. Among compost nerds, South Korea is considered nirvana.

We reach the end of this field of rows, where a large trommel machine – an industrial sieve, effectively – has been set up. 'We do two grades: a 25mm, which is what farmers will take and dig into their fields; then we do a 10mm, which is the stuff that we bag up and sell.' The trommel screen removes contaminants. Often, that will be large organic material that has yet to decompose – wood chunks, errant stones, etc. But the biggest problem is plastics. Since 2017, composters have seen a rapid influx of so-called 'compostable' or 'biodegradable' plastics, sales of which have soared, as an alternative to single-use plastics. Pretty much every independent café near where I live has switched their plastic takeaway containers to 'compostable' alternatives. There's one problem: for the most part, these plastics do not actually decompose. 'They're not biodegradable,' James says. 'They need exposure to sunlight and moisture and even if you've got them in the middle in all that heat, they're unlikely to break down in seven weeks.' In Cumberlows' case, plastics make 'maybe five per cent' of the total waste collected, but removing it makes a sizeable dent in the bottom line. 'The skips to clean it up cost me a fortune – ten grand a month. You're paying for that to go away.' As a result, it has become common practice for compost facilities to screen out these 'biodegradable' plastics, which are burnt instead.

The idea of bioplastics is as old as plastics themselves – the very first plastic, Parkesine, was a cellulose derived from plants. Today, co-opted by greenwashing and corporate interests, the term has come to represent a kind of mythic solution to the plastics crisis. Imagine! A material with the miraculous properties of plastic that would simply moulder away like leaf litter, dissolving benignly into the soil. In reality, bioplastics – loosely defined as plastics made from organic matter – come in several forms. *Bio-based* plastics are made from feedstocks such as corn and sugarcane, but are functionally the same as those made from fossil fuels. Coca-Cola, PepsiCo and Suntory[20] have all used bio-based PET in their bottles. These plastics are often marketed as more sustainable, but are functionally identical to fossil fuel-based PET.

Oxo-biodegradable plastics are ordinary polymers – PET, HDPE, LDPE – containing an additive such as metal salts, which react with oxygen to make the plastic break down more quickly. These plastics have been around since at least the 1980s, and are sometimes marketed as direct replacements for single-use items such as plastic bags. However, recent evidence has shown that these plastics don't decompose so much as break up into smaller pieces, leaving behind micro- and nanoplastic particles.[21] As a result, both the UK and EU are currently considering banning the materials entirely.[22]

Then there are *compostable* plastics. When I first heard the term I, like most people, presumed that meant the plastic would break down inside any compost heap. Not so. According to the official (UK and EU) definition, 'compostable' refers to plastics that break down under *commercial* composting conditions: twelve weeks inside a high-temperature commercial composting facility or equivalent anaerobic digester.[23] The most widespread of these is Polylactic Acid (PLA), a plastic often made from sugar cane. PLA can be used for everything from bottles to coffee-cup liners to single-use cutlery, and has accordingly taken off within the food and drink industry.[24] (To add more confusion, not all compostable plastics are bioplastics. Polybutylene adipate terephthalate – PBAT – one of the plastics industry's great new hopes, is made from virgin petrochemicals.)

Most people I have met within the waste industry hate compostable plastics. In the UK, there is no dedicated collection for compostables, meaning that they either end up inside compost bins (where they don't decompose) or in the regular trash. Clear PLA bottles mixed up on a conveyor belt are easily confused with regular PET, and therefore contaminate recycling streams. Composters, meanwhile, remove them because they don't degrade properly, and PLA and other 'compostable' bags can clog machinery, shutting down lines. The result is that compostable plastics are often more greenwashing than reality. For example, in 2018, the UK Parliament switched its single-use plastics in the House of Commons and House of Lords to compostable materials by the British company Vegware. However, an investigation by *Footprint* found that due to problems with collection, most of the waste was incinerated instead.[25]

Imogen Napper and Richard Thompson, two academics at the University of Plymouth, have been studying biodegradable plastics for years. In 2016, Napper submerged five types of plastic bags – biodegradable, oxo-biodegradable, compostable, and HDPE – into Plymouth harbour, exposed others to the air, and buried others in the soil, and left them there for three years. The compostable bags disappeared in three months underwater but persisted in the soil for twenty-seven months. One biodegradable bag was still intact six years later – and is in fact still intact even now.[26]

Danielle Purkiss, a materials scientist at University College London's Plastic Waste Innovation Hub, studies how the claims made by compostable packaging manufacturers stack up with reality. 'The problem is when you start applying a term like biodegradability to an object it's meaningless,' Danielle says cheerfully. We're speaking over a video call – the UK is in the midst of a coronavirus wave, and so entry to her lab is limited. Danielle has stylish glasses and the cheerful, engaged manner of someone truly in love with their subject. The major problem, she explains, is that words like *compostable* and *biodegradable* lack context. 'A biodegradable newspaper might end up in the water, and might end up in soil, and it might end up in the open air. Those are incredibly different environments, which means that testing for that is potentially impossible,' Danielle says. 'But legally there's really nothing to limit the use of the word biodegradable or compostable as a kind of descriptor – and that is massively misleading.'

Currently, legal standards for compostable and biodegradable plastics are set on fixed criteria, tested in controlled laboratory settings. But laboratory settings rarely reflect the real world. 'We've been looking at how people actually compost and – surprise, surprise – people don't compost in the way that the lab tests work.'

Like soil, compost has *terroir*. Biodegradation varies according to a panoply of both human and environmental factors: temperature, humidity, wind, local flora and fauna. And then there are human factors. How often is it turned? What kind of storage is being used? To try to get at some of these questions, in 2019 Danielle and her colleague Mark Miodownik launched the Big Compost Experiment, a nationwide citizen science project to examine the biodegradability of compostable plastics at home. Anyone with a home compost bin can sign up to the website, where they're asked to record themselves composting one of these plastics, to see the results. 'It's been super interesting and very fun, having a little peek inside everybody's compost bins and in their gardens,' Danielle says. (She really loves compost.) I sign up myself, cheerfully snapping shots of Vegware frappe cups going into my bins. Although Danielle's lab is yet to publish its results, the early signs are not good for compostable plastics. According to Miodownik, more than 60 per cent of plastic samples logged so far do not seem to biodegrade.

The results can be depressing. 'We were sort of joking earlier that everyone who works in this research group comes away feeling slightly damaged,' Danielle says. 'You see and understand these things that you blissfully didn't have to deal with before, but now can't stop thinking about.' I have felt this, too: the way that waste takes hold of you without realising, like an earworm. It can make you cynical; over time, the sheer scale of the problem, and the greenwashing associated with it, take a toll. 'It's definitely changed our outlook on these things,' she continues. 'We're also a lot more cynical about stuff that claims to be green.'

The joy of compost, however, is that it can counter the sense of doom. I'd seen it with John and James, and now with

Danielle, too. 'I think fundamentally there's a very human aspect to it. It's about living things. It's a nice way to understand that you're connected to everything,' she says. I had felt it myself, the strange pride of seeing something turn from waste and decay to a teeming ecosystem, full of life. A resurrection of sorts; the cycle beginning again. It's a bit like the wonder you feel when you watch a seedling sprout, only greater. As I started to compost, my sense of guilt about waste began to lift.

For all of compost's charms, most food waste in Britain is not composted, but digested. Industrial anaerobic digesters (known internationally by their major product, biogas) act less like a compost pile, and more like the inside of a stomach: the organic waste is mixed with water inside vats full of methanogenic bacteria, which are stirred and heated to encourage anaerobic breakdown. As inside a landfill, this results in the production of biogas – a mixture of methane and carbon dioxide – which can be siphoned off and burned to generate electricity. The remaining solids, digestate, can be spread on fields as fertiliser.

In recent years, the digestion business has soared. There are now more than 579 anaerobic digestion plants in the UK.[27] Since 2015, the amount of electricity produced by AD plants in the UK has doubled. Most of those are fed by farms, which increasingly use digesters to process manure and crop waste. Supermarkets and large food manufacturers have also seen an opportunity in anaerobic digestion, and have begun sending large amounts of their food waste there to turn into cheap electricity. This helps supermarkets and manufacturers both with their zero waste-to-landfill target, and also to make claims about reducing emissions.

There is an anaerobic digester a few miles from my house, run by the energy company Biogen. It's a quiet, unassuming facility. Most people don't really notice it, except on the days that digestate is being spread on the fields. Then the smell can travel for miles, like manure spreading, but worse. I try to think of it as a good thing – at least that food isn't in landfill – but it doesn't hold off the aroma for long.

Biogas is a very old process: London was using it to fire street lights in the nineteenth century.[28] In rural China, tens of millions of rural homes are fitted with small-scale anaerobic digesters, which tenants use for cooking and lighting and are fed with farm and animal waste. In Sweden, some public buses and taxis run on biogas generated from municipal waste.[29] To proponents, it is a sustainable source of green energy, a way to harvest something usable from our waste while also generating fewer carbon emissions than landfill. Critics, however, argue that biogas is an inherently inefficient way to generate electricity. (As Tristram Stuart has written, digesting a tomato generates a fraction of the energy used to grow and transport that tomato.)[30] The better thing for the environment is always to waste less food. And yet, what was designed to process waste is now generating more of it. Because biogas generators pay for their feedstock, farmers are increasingly growing crops such as maize specifically to be digested. In 2019, 1.9 million tonnes of crops were grown in the UK purely as biogas feedstock. In agricultural terms, that's enough to cover 440km², an area larger than the county of Rutland.[31] In Germany, whose 9,000-plus digesters are aided by government subsidies, the rate is even higher; some critics suggest that the amount of farmland now being used to grow biogas feedstocks has increased the price of food.[32] Biofuels have gone from something designed to consume waste, to something that generates it: while people go hungry, we're growing food just to throw it away.

* * *

Over time, policymakers and academics have developed a hierarchy for disposing of food waste: feed food waste to animals, send the manure to digesters, and compost the sludge that remains. This orthodoxy strikes me as a sensible way to perpetuate our food system, if not to reduce food waste. (If we're interested in saving the planet, then we should be growing less food to begin with, livestock included.)

Rather than treating food waste and sewage as separate issues, there is also a growing movement that urges us to once again see them as, if not quite the same, then a problem with a unified solution. By breaking down our waste, the theory goes, we can solve our waste problem and return to the kind of natural closed cycles that nature intended, rather than having to burn or bury it. This, according to some proponents, could mean returning to the days of nightsoil, at a time when all organic waste – human, plant and animal – returned to the field.

In truth, we are already there. Mogden sewage plant, for example, sells its settled sludge as fertiliser, as do many water companies throughout the UK. In the US you can buy Milorganite, a fertiliser produced using the city of Milwaukee's sewage sludge, in many garden centres (it is said to be particularly good for golf courses). Sewage sludge itself is commonly composted, in order to destroy any remaining pathogens and stabilise volatile compounds. EU countries apply around 6 million tonnes of biosolids to farmland every year, the UK another 3 million tonnes.[33]

Composting toilets, too, are no longer just for music festivals and the eco-conscious, like Compost John. 'Ecological sanitation' (or EcoSan) toilets are becoming common in the developing world, from Haiti to Kenya to India. In poorer countries, these systems have value: not only do they not require a constant stream of running water, they are cheap to build, and provide a free and regular supply of fertiliser. It's not just number twos, either. The burgeoning 'peecycling' movement argues that we should all be collecting our urine – pee being rich in nitrogen and phosphorus – and using it to feed plants, rather than washing it down the drain. This is more than just about reducing waste. The nitrogen in urine is seen as a more sustainable source of nitrogen than synthetic fertilisers; phosphorus, one of the most essential building blocks of life and essential to plant growth, is a dwindling natural resource.[34] A recent pilot study in Niger found that women who fertilised their pearl millet crop with urine saw their yields increase by 30 per cent.[35] Tanum, a town in Sweden, requires all new buildings to have urine-recycling toilets; in Paris, 600 new 'eco' apartments are currently under construction with pee-cycling toilets, whose contents will help to fertilise plants around the city.[36]

These are in a way a belated recognition of the material and mineral value in our waste. In 2019, the fertiliser giant Yara and the waste conglomerate Veolia signed a deal to effectively mine phosphorus from farm and food waste, through the use of compost and anaerobic digestion.[37] Germany and Switzerland have both passed laws making phosphorus recycling from sewage sludge and slaughterhouse waste mandatory, in an attempt to both recover the phosphorus and prevent it from polluting rivers.

As I was learning about these practices, I couldn't help but think back to Edwin Chadwick, whose vision for the city of London was akin to a circulatory system, in which clean water would be carried in via pipes, and then wastewater back out of the city, and onto farms. Chadwick's vision was to 'complete the circle, and realise the Egyptian type of eternity by bringing as it were the serpent's tail into the serpent's mouth.'*

It's an appealing ideal, to return to natural cycles of breakdown and rebirth. The circle, however, is harder to square than we might think. Recent studies have shown that sewage sludge spread on farmland can contain a number of chemical toxicants, including per- and polyfluoroalkyl substances (PFASs), dioxins, furans and PCBs.[38] A recent study by the Environmental Working Group, a US non-profit, estimated that as many as 20 million acres of US farmland may already be contaminated.[39] Sewage plants are designed to filter out microplastics – but using sludge as fertiliser dumps these plastics onto farmland, which can be taken up by animals, and washed into our rivers. In 2021, a research group in Sri Lanka found that compost was also a 'prime source' of microplastics found in surface soils; other toxic chemicals, such as metals and PFASs,[40] can bind to these plastics in the soil.[41] The effect on soil ecosystems is as yet poorly understood. It is unlikely to be good.

This is the uncomfortable truth we must acknowledge when we talk about returning to a natural system of rot and decay. As much as I have grown to love compost, and admire its potential in building a healthier waste system, any hope of a circular, pastoral idyll is at best naive. We have not only broken the nutrient cycle and cleaved open the metabolic rift; the world we have created in its stead is a toxic one, ridden with thousands of manmade substances of unknown fates and harms. By returning our waste to the earth, we are now administering poisons of our own making.

Before we can close any loops, we need to go on a detox.

---

* I have never understood this symbol. The ouroboros, as it's called, is commonly used to signify an eternal loop, but in literal terms would mean the opposite. The snake would bleed to death, and in any case could not eat its own head, which would be left behind as waste. I suppose it's a more apt metaphor for our waste system that way.

PART THREE

TOXIC

9

UNHOLY WATER

'You would only know after you have eaten it'

On winter days, the river in New Delhi foams like a bubble bath.

It's a cold January day, and I'm standing on the bank of the Yamuna, the river that bisects the city from north to south. I'd read about the Yamuna foaming in the winter, but wanted to see it for myself, so I've come to the Okhla barrage, a weir built in 1987 to control the water's flow. The sky is particulate grey, the roads choked with traffic, but sure enough: the river is bright white and thick with foam. Suds whipped into airy peaks blanket the surface, colliding and breaking apart like a collapsing ice floe. Lather piles high on the water's edge. Spindrift clouds thick as candy floss catch the wind and float free, landing on my skin.

It's an uncanny thing to witness. Confined to the sinks and basins of our porcelain-enamelled bathrooms, bubbles usually indicate cleanliness, purity. But out here under the open sky, their presence indicates the opposite: contamination. Rapids might bubble and waves may froth, but we all know, almost intuitively, that rivers are not supposed to behave like this. I scrape a fingerful of foam off my sleeve, thick and weightless. Looking at it, I'm immediately struck by a gut feeling of violation – that something is wrong.

Something *is* wrong. The foam is caused by pollution: surfactants and phosphates discharged in untreated sewage and in wastewater from textile factories and paper mills upstream.[1] Surfactants are chemicals designed to reduce the surface tension of liquids (like politicians, they make things slipperier) and are found in, among other things, soaps, detergents, paint, adhesives, and cosmetics. Although not considered harmful in small concentrations, in large quantities the chemicals can cause skin and respiratory problems, and disrupt aquatic ecosystems.[2] Crucially, surfactants foam – and so as the water cascades through the barrage, a once pristine river is churned up into something resembling a toxic cappuccino.

To Hindus, the Yamuna itself is more than a river. 'She' is a goddess, the daughter of the sun god Surya and the cloud goddess Sanjna, and the twin sister of death. Rivers are sacred in Hinduism, the source of both spiritual and physical cleanliness, and river water is central to ritual and prayer. Every year millions of Indians flock to the riverbanks during religious festivals. The Kumbh Mela, a bathing ceremony celebrated every twelve years, is considered the largest gathering of human beings on Earth.

Of seven sacred rivers in India, the Yamuna's holiness is second only to the Ganges. In Delhi, it is crucial to daily life, providing the capital with 70 per cent of its water supply. More than 57 million people are estimated to rely on it for drinking and bathing. However, as India's economy has boomed, heavy industries have taken root along its banks – sugar, tanneries, distilleries, electronics, leather, plastics – which discharge their untreated wastewater into the river. Consequently, studies have found that the Yamuna is contaminated with unsafe levels of heavy metals, including cadmium, chromium, and lead.[3] Then there are the sewage outfalls, which release 477 million litres of raw effluent per day directly into its waters.[4] The combination is foul. In this stretch of the river, the Indian government's Central Pollution Control Board has measured fecal coliform bacteria at 943 times the state-recommended concentration, and faecal *streptococci* (used as an indicator of dangerous pathogens) at 10,800 times the recommended level.[5] As a result, the Yamuna is no longer considered safe for drinking or bathing.[6] In 2017, it was declared 'biologically dead' – that is, unfit to support[7] aquatic life.[8]

This particular morning, the banks are quiet. Waste lines the sandy foreshore: discarded flip-flops, polystyrene, prayer beads, a single Nike shoe, a Huggies nappy. Relics of celebration and mourning. In the shallows, someone has placed an idol of a praying man – a deity, perhaps – although I don't recognise it, its face half-obscured in the foam. Below the barrage, suds pile against the *ghats*, steps that lead into the water used for ritual and prayer. It's there that I first see the diver: a man half-submerged in the toxic water, carrying a bag of human ashes.

Hindus cremate their dead, in devout and familial ceremonies.[9] After death, the deceased's body must be washed by a family member – usually the eldest son – and burned on a wooden pyre. The ashes of the deceased are then scattered into moving water, so that their spirit might be purified and carried off to the next life. These ceremonies ideally take place on the Ganges, at the holy city of Varanasi. However, not everyone can afford to make the trip; the Yamuna, as a tributary of the Ganges, is a suitable alternative.

Three men, who I take to be the family of the deceased, watch from the *ghat* as the diver empties the plastic bag into the turbid water. It is not a dignified farewell. The cremains swirl and clump on the surface, dispersing slowly; a few catch the wind and are blown onto the foam. After a few words of prayer, the diver rinses off his hands and body, climbs out, and dresses in a sand-coloured jacket and trousers. The family pay him and leave, clutching the empty plastic bag.

I introduce myself. The diver's name is Feroz Napa, 'although everyone calls me Napa the diver,' he says. A wiry man in his later years, Feroz has high receding hair and a thick silver beard, his skin still wet beneath his clothes. Feroz has lived near the Yamuna since he was a boy. 'I remember twenty years ago this water used to be really clean,' he tells me. But that started changing, he says, when the factories began to spring up in Wazirabad, upstream.

Once the river started foaming, Feroz explains, fewer and fewer people were willing to perform the sacred rites for their deceased. 'They are scared of the water.' That's where he comes in. For a small donation – families typically pay anywhere from 500 to 2,000 rupees (£5–20) – he will surrender your loved ones to the afterlife, while you watch from the safety of the bank. Every morning, Feroz rises at 4 a.m. and comes to the river, staying until the sun sets. During the heights of the Covid pandemic – thought to have killed as many as 4.7 million people in India, the highest total of any country – Feroz was performing the rites for as many as fifty families per day. 'I used to be drunk all the time,' he says. 'I would take a glass of alcohol, go in there, dump the ashes, then I would come back and drink another glass.'

When he isn't performing funerals, Feroz dives for valuables. Supplicants sometimes drop money or jewellery during prayer ceremonies, which he can scavenge to sell. On rarer occasions, he finds bodies. We've not been talking for long when he pulls out his phone and, unprompted, shows me a video of a teenage boy with mangled limbs, lying on the steps a few feet from where we're standing – a jumper. The road bridges that span the river over the barrage are, unfortunately, a popular location for suicide attempts. 'Maybe twice a month,' Feroz says, grimly. The boy survived, but he never saw him again.

The Yamuna was not always like this. Feroz remembers a time, decades ago, when its waters were clean and bountiful. 'A thousand species of fish used to live here,' he says. 'Most of them died. Now only one or two species are able to survive.' Those that do, mostly Clarias catfish and Nile tilapia, are invasive species, able to handle the impurities in the water. In hot weather, excess nutrients from sewage can trigger algae to bloom. When this happens, the dissolved oxygen levels in the river collapse, triggering mass fish die-offs.[10] 'This happens every year,' he says. On one recent such occasion so many fish carcasses washed up on the banks that the city hired labourers, Feroz among them, to collect the dead remains. Rather than dispose of the fish, the workers sold them to the markets at a discounted rate. He sees my horrified expression. 'There is no issue for us in selling. You would only know after you have eaten it.' The polluted water dulls the flavour, he explains. 'It's less tasty. Sometimes the flesh goes black, instead of pink. A person who knows the taste would know that there is something wrong.' (Needless to say: if you're ever in Delhi, think twice before you order fish.)

I ask Feroz if he's concerned about the pollution. He shrugs. 'I'm used to it,' he says. 'I'm not scared.' This is not even the river at its worst, he explains. 'In the summer months, the water flow stops and the river becomes stagnant. The water turns black, and it smells awful.' Feroz, like many Hindus, believes that the water of India's holy rivers are incorruptible. 'If a person believes in the holiness, he doesn't care about the pollution,' he says. His belief that the pollution won't harm him is an act of faith.

* * *

In truth, I hadn't come to India to write about the Yamuna. I was more interested in the Ganges.

The Ganges basin is one of the most densely populated river basins in the world. Home to close to half of India's population,[11] it provides water and irrigation for an estimated 600 million people.[12] In Hinduism the Ganges is the mother river, whose waters were formed, according to the *Bhagavata Purana*, when Lord Vishnu himself punctured a hole in the universe with his toe and divine water flooded into the world.[13] Accordingly, Hindus hold that the Ganges contains divine properties. The Mogul emperor Akbar called it 'the water of immortality'.[14] In the *Ramayana*, Lord Vishnu asserts that man's sins are destroyed by consuming it, touching it, or even looking at it. Ganges water is widely used in Hindu prayer and ceremony, and is its own pocket industry: you can buy plastic bottles of *Ganga jal* from stalls all over the subcontinent, or order one on Amazon for as little as £3.

And yet, despite its sacred status, the Ganges is widely considered one of the most contaminated rivers in the world. The UN has called it 'woefully polluted'.[15] Much of this is due to India's sanitation problem: in 2016, according to one report, as much as 78 per cent of sewage in India remained untreated. With nowhere else to go, effluent ends up in rivers and streams[16] – around 4.8 billion litres of it every day.[17] But the Ganges' other pollutants are manifold: industrial waste, agricultural runoff, pharmaceutical pollution and, inevitably, plastics. If any river manifests our global problem with waste and rivers, it is this one.

How is it possible, that perhaps the most sacred river on the planet could end up among its most polluted? I'm not the only one asking. In 2014, the newly elected Prime Minister of India, Narendra Modi, announced arguably the most ambitious element of his government's 'Clean India' campaign: the Namami Gange project, an expansive infrastructure plan to finally clean up the country's holiest river. Modi is not the first to have made such an attempt. India's government has spent fortunes[18] in recent decades trying to fix the river's pollution problems, but every effort has been mired in poor delivery and corruption.[19] Namami Gange is the largest attempt to date, comprising more than 340 individual engineering projects, including miles of new sewer lines and dozens of new sewage treatment plants. At the same time, Modi's controversial government has built 110 million new latrines across the country, in an attempt to end the once-widespread practice of open defecation. Combined, the effort might represent the most ambitious sanitary engineering programme since Bazalgette built London's sewers. If I wanted to understand the ongoing relationship between our waste and our environment, where better than in the midst of the biggest cleanup operation in the world?

If we're telling the story of waste in our rivers, we should start with the time that the Cuyahoga river caught fire. I remember being stunned by that image the first time I read about it. *The river caught fire.* But in fact by 1969, the year of this particular fire, the Cuyahoga, in Ohio, had been regularly catching fire for over a century: 1868, 1883, 1887, 1912, 1922 – well, you get the picture.[20] By the 1950s, heavy industry had reduced many rivers in the Global North to open sewers. Waterways, thick with sewage, oil, and industrial runoff, routinely stank. Fish populations plummeted. In the United States, several major rivers were so polluted as to be virtually dead. By 1969, the Cuyahoga was a fetid stew. The culprit was industrial waste pouring forth from factories along the riverbanks from

Cleveland to Akron.[21] 'Rings of oil circled on its surface like grease in soup,' as one witness described at the time.[22] The result was water that was more than toxic; it was flammable. The Cuyahoga was far from the only river to ignite: at various times during the late nineteenth and early twentieth centuries, the cities of San Francisco, Baltimore, Philadelphia, Buffalo and Galveston all witnessed their waters set ablaze. By the Cuyahoga's standards, the 1969 blaze was relatively small, lasting just twenty-four minutes before it was put out. But the fire caught the imagination of the American public – then in the throes of the early environmental movement – and helped to spur the creation of both the US Environmental Protection Agency and the Clean Water Act.

There's an old saying in environmental science: 'the solution to pollution is dilution.'[23] The origins of this slogan are obscure, but the idea behind it is ancient. In Greek myth, when Eurystheus gives Heracles (or Hercules, to both Romans and Disney fans) the odious task of cleaning out the Augean stables, the hero does so by redirecting two nearby rivers to flush away the filth. For as long as we've had waste to dispose of, we've relied on water to wash it clean – or at least to convey it out of sight. It is only a slight oversimplification to say that this attitude describes our regulatory approach to pollution for most of human history, and lingers even now.

Like any good myth, the dilution maxim is rooted in truth. Left to their own devices, rivers and oceans are to a certain extent self-purifying: microbes in the water column break down organic matter, which are feasted on by algae. Micro- and macrophages chomp on pathogens. Root systems, grasses, and other plant life filter and sequester certain pollutants, even metals. (Some plants are so effective at this that there exists an entire field, phytoremediation, dedicated to using plants to clean up environmental disasters.) Over time, ecosystems tend towards a kind of dynamic equilibrium – but upset that balance with, say, a massive surge of anthropogenic waste, and a tipping point is crossed.

The dilution maxim is tied to another: that 'the dose makes the poison'. This observation, credited to the Swiss physician-philosopher Paracelsus, is the fundamental precept of toxicology. Two paracetamol will ease a hangover, 200 might kill you. Applied to nature, this concept is sometimes called assimilative capacity[24] – how much pollution you can flush into an ecosystem until the ecosystem (or, say, a human body) can no longer handle it. It is by this guiding principle that governments have for decades set permissible limits for various pollutants in drinking water.[25] In the UK, for example, drinking water may contain a maximum of 50 micrograms per litre of cyanide, 5 micrograms of cadmium and 0.030 micrograms of the pesticides aldrin and dieldrin, both banned carcinogens.[26] Any amount below these permissible limits are considered safe. The geographer and waste academic Max Liboiron calls this 'threshold theory'.[27]

Threshold theory is replete with problems. For one, it assumes what Liboiron describes as a colonial relationship with land, ecosystems, and indigenous populations; chemical wastes pumped into a river may poison those downstream, for example. It doesn't account for cumulative impacts – all of the ways by which we are exposed to chemicals, whether through our food, skin and the air – nor the thorny issue of how chemicals interact and combine both in nature, and within the body. It is politically biased: regulators often ignore certain economically advantageous pollutants, such as nitrogen and phosphorus-based fertilisers, the runoff from which have contributed to the creation of oxygen-starved dead zones in rivers and oceans worldwide. Moreover, as Liboiron will tell you, the fundamental principle of threshold theory is not to *eliminate* pollution so much as to simply dictate how much pollution is allowed, where, and by whom.

Then there's the tricky issue of calculating the limits themselves. Many of the 'safe' thresholds we do have, such as those for lead and mercury, have been periodically revised downwards over time as new evidence of their toxicity has arisen;[28] in 2021, for example, the European Food Safety Authority proposed new safety standards for Bisphenol-A that are 100,000 times lower than the previous limit.[29]

Environmental regulators historically worked on the assumption of safety – that is, chemicals would be approved unless there was incontrovertible evidence that they caused harm. Infamously, when the US government passed its Toxic Substances Control Act in 1976, it gave a free pass to around 62,000 chemicals already in circulation, meaning most were never assessed for toxicity. (That law was finally amended in 2016.) Today, around 86,600 chemicals are registered with the EPA, of which it claims around 42,000 are 'active' in the marketplace. Of those, the US government has gathered sufficient evidence to ban or seriously restrict just fourteen.* (The International Agency for Research on Cancer, for comparison, currently lists 214 chemicals as 'carcinogenic' or 'probably carcinogenic' to humans, and another 320 'possible' carcinogens.)[30] Proving the toxicity of a chemical is an expensive and laborious process. We are exposed to thousands of chemicals every day. How can you measure the effect of just one? Moreover, when substances *are* banned, the chemical industry has a habit of

---

* These include mercury, asbestos, lead, dioxins, PCBs, chlorofluorocarbons, formaldehyde, certain metalworking fluids, and hexavalent chromium, on which more in a moment.

substituting them with nearly identical compounds that later turn out to be toxic as well – replacing DDT-era pesticides with bee-killing neonicotinoids, for example.

The old assume-everything-is-safe approach is changing – slowly. In 2006, the EU passed a law known as REACH, which requires companies operating in Europe to submit safety data for both new and existing chemicals in mass production. About 23,000 chemicals were registered,[31] of which the EU currently lists 224 as 'substances of very high concern'. It has not been a simple process. REACH has been criticised for being slow and ineffective, while a 2019 review by the European Environmental Bureau found that of the 2,000 safety dossiers assessed so far, 70% were not compliant with legal requirements,[32] and 64 per cent of the chemical companies' submissions 'lacked the information to demonstrate the safety of the chemicals marketed'.[33]

How many chemicals are swirling around in our environment? The reality is we don't really know.[34] Unless a substance is known to be toxic, we tend not to test for it. One recent study, which attempted to comb twenty-two existing chemical inventories from around the world, came up with the figure of 350,000 known chemicals currently registered for production and use globally, a figure rising fast.[35] In fact, some scientists argue that we have already passed the 'planetary threshold' for chemical pollutants,[36] at which point the number of toxic substances in circulation start to threaten the integrity of planet-wide ecological processes.[37] We have not just poisoned the well, we're now swimming in it.

Ride an express train east from New Delhi for five hours, across the lush fields of Uttar Pradesh, and you'll arrive in Kanpur, on the banks of the Ganges. Kanpur is an industrial city, with a rich history of manufacturing and trade.[38] As early as the

eighteenth century, the colonising British East India Company
set up a trading post here, using Kanpur's convenient position
on the river to better exploit (see also: steal) India's natural
wealth.[39] In the 1860s, the occupiers established a leather
factory to produce saddles and boots for the British army. The
trade spread into a thriving and filthy manufacturing sector,
pumping out leather and textiles across the world, and earning
Kanpur the nickname 'the Manchester of India'. (This term
dating from when Manchester in England was known for its
textile mills, rather than its ludicrously rich football clubs.)

At its peak, more than 400 leather tanneries operated in
Kanpur and its surrounds, producing as much as a third of
India's £2.5 billion leather industry.[40] Even now, you can't walk
far along its streets without seeing market stalls bearing the
products of workshops and factories across the city: sandals,
belts, handbags, jackets, heels, boots. When I arrive in mid-
January, a cold front has settled over the city, and the wind is
thick with the smell of car exhausts and unboxed shoes. Smog
hangs in the air, filling my mouth and infiltrating my clothing.
In 2016, the WHO ranked Kanpur as the city with the worst
air pollution in the world.[41] When I check an app on my phone,
the particulate levels are twenty-nine times the safe limit.[42]
Suddenly I'm glad to be wearing a mask.

Tanning leather has always been a noxious process. First,
livestock must be slaughtered and skinned. The hides are then
washed, scoured, and bathed in acid salts to kill off any remaining bacteria or mould. In pre-industrial times, this would have
been done with the aid of urine, animal faeces, and even brains;
later processes involved a variety of toxic chemicals, including
formaldehyde and arsenic, all of which need to be dumped
somewhere, almost always in rivers. The stench arising from
tanneries meant that as early as the Middle Ages, tanners were

pushed to the outskirts of cities and downstream, lest their waste sully the supply of drinking water.[43] Around 75 per cent of the hide is wasted in the tanning process; by one measure, the leather industry produces 200 times more waste than actual leather products.[44] The tanning process itself is typically done in one of two ways: vegetable tanning, which involves soaking the hides with tree bark or vegetable oils; or chromium tanning, which bathes the hides in chromium sulphate. Vegetable tanning is less polluting, slower, and more expensive – thus, about 85 per cent of the world's leather is currently produced by chrome tanning.[45]

Chromium sulphate* is a relatively harmless chemical, thought to play a role in human metabolism. (You wouldn't want to bathe in it, or anything.) But the problem is that chromium tanning can produce chromium-6, or hexavalent chromium, as a waste by-product. Chromium-6 is a potent carcinogen, exposure to which is linked to lung, stomach, prostate, bladder, kidney, bone, and thyroid cancer.[46] (You may know it as the chemical made infamous in the United States by Erin Brockovich.) Leftover chromium-6 can remain in leather scraps and in dust, which collects in wastewater during the tanning and rinsing process. In Kanpur, the tanneries emit tens of millions of litres of effluent into the Ganges every day.

Something they don't tell you in a shoe shop: when leather has been first been tanned with chromium salts, it comes out the colour of a cloudless sky. At this stage, the hides are known as 'wet blue', and in Kanpur, they are everywhere. I want to understand where the wastewater into the tanneries is coming from, so along with my producer Rahul Singh – a local filmmaker acting as my fixer and translator – I have come

---

* Chromium (III) sulphate, or trivalent chromium, to get technical.

to Jajmau, the historic hub for the leather trade, downriver of the city centre. Jajmau is an old neighbourhood, its narrow lanes edged by overflowing gutters. Butchers along the roadside hack at bloody joints, their chickens pecking at each other inside cramped cages. Vegetable sellers pull carts of onions and chilli peppers, jamming the roads. Most jarring are the chemical shops, advertising various salts and acids in bright azure barrels. Most of the businesses here exist to support the tanneries, which dominate every street corner. Outside each is a mandatory 'hazardous and chemical waste' notice, listing the name of each business and the chemicals being used therein: chrome, sulphuric acid, formic acid, dyes. Most of the signs are faded, and none up to date.

My arrival in Kanpur, as it transpires, is poorly timed. It's the week of Makar Sankranti, a Hindu festival that begins Magh Mela, a forty-seven-day period of religious celebrations. Many Hindus mark the occasion by taking ceremonial baths in the Ganges; as a result, the government of Uttar Pradesh has forced the tanneries to close, in order to lower the level of pollution in the river. Luckily for me, many tannery owners have ignored the edict. Tanners pass with pickups laden with bright blue hides. Used bags of chemicals linger empty in the road.

On one street, we duck through an open doorway and into a small workshop. It's a cramped space, barely lit, with a concrete floor and bare brickwork. Hides in great heaps surround us. The owner, a thin man with a wiry moustache and bare feet called Abdul Hazam, is working on a machine grinding the hides into suede. Leather hides are normally split – the thick, smoother outer hides used for higher grade leathers (so called 'full-grain'), and the inner half used for thinner leathers, including suede. Hazam buys hides in bulk – from ₹5 to ₹25 (5p to 25p), depending on the quality – buffs them into suede, then sells them on to the larger tanneries for a profit. Kanpur is set up in this way: there is a neighbourhood that does wet treatment, another for dry, others for finishing and manufacturing – an assembly line in city form. Abdul works on the grinding machine for twelve hours a day. 'If I work for one week, I can get at least ₹2,000 [about £20],' he says. It's a dangerous task. 'If you make a single mistake, your hand will go inside,' he says, gesturing towards the belt mechanism. People lose their hands.

Over the last decade, the government of Uttar Pradesh has issued numerous executive orders to try to limit the chemical pollution emitted by Kanpur's tanneries. In 2020, it ruled that only half the tanneries may operate at once, alternating every fifteen days. For the tanneries, the move has been economically devastating, cutting revenue by half, and resulting in a spike in unemployment. A million people once worked in the leather trade in Kanpur; now, with the leather trade slowed, the pavements are lined with people waiting for work.[47]

The chromium effluent from Kanpur's tanneries is supposed to be treated at a treatment plant in Jajmau. Built in 1986 under a previous effort to clean the Ganges, the plant is designed to remove the chromium from the water, recover it into reusable salts, and discharge clean wastewater into the river. The tanneries have to pay for their effluent to be treated, and to buy back the recycled chromium at what several tanners say are unreasonable rates – so not all tanneries follow the rules. Worse, the treatment plant is now old and degraded. Several tannery owners told me that often it does not even operate, and when it does, the plant's capacity of 9 million litres per day is a small fraction of the tanneries' total output, estimated at 40 million litres a day.

The city's sanitation officials are ignoring my calls, so the next morning we take a taxi to the treatment plant in the hope of talking to someone in charge. When we arrive, the front gate is open, but the plant appears deserted. I wander tentatively inside to try to find someone. The buildings are overgrown, the concrete stained and cracked. Inside, the wastewater is passed through an ancient-looking series of pools to be treated. A blueish-grey crust has formed on the surface; it looks like it has not operated in days. We head back outside. Across the road, someone has placed hundreds of blue hides out in the dirt to dry in the sun. There are cows wandering among the hides, picking at grass scraps, hopefully oblivious to the dead beneath their feet.

The largest tanneries are no longer in Jajmau, but have moved to Unnao, an industrial city about 10 km north-east, where a handful of factories have set up on a soulless industrial park. When we arrive, it is also deserted, the tanneries hidden behind high walls. The streets are neatly arranged in a grid, and the open gutters that line the roads are clogged with garbage and cerulean-blue sludge. Unlike in Jajmau, these mostly appear to be adhering to the festival restrictions. Thankfully, one tannery owner has agreed to meet me.

Asad Iraqi is a smartly kept man with a neat grey beard, side parting and spectacles, a purple shirt peeking out from his gilet. A thirty-year veteran of Kanpur's leather industry, Asad is the general secretary of the Leather Industry Welfare Association, which represents the tanneries' interests both locally and internationally. When we arrive at his tannery, Aki India Industries – a neat, modern building on the edge of the industrial area – we are welcomed upstairs into a spacious air-conditioned boardroom. This is what Asad modestly describes as a 'medium sized tannery', processing 1,000 hides per day. (The largest, he explains, can process 2,000 a day.) Swatches hang on racks in a dazzling array of textures and colours. 'These are the fine leathers we are making,' Asad explains. 'For bags, for shoes. We are making them for brands in the UK.'

We've barely been talking for a minute when he says: 'You want to see the drain?' He leads me back out into the street, and a few yards away to a small road bridge across a nearby *nullah*, or waterway, in reality little more than an open sewer. As I'd seen elsewhere, the water is near black, and coated with plastic waste. 'A lot of untreated garbage is brought here from the city. How can you blame the industry, when the industry starts here?'

Asad is keen to prove that not all tanneries are polluters. We head down to the production floor, where what must be tens of thousands of hides are heaped in front of the tanning drums, massive rotating chambers big enough to hold a pickup truck. A few workers are lingering about the place, moving stock, fixing up machinery. Out back, Asad shows me the plant's chromium treatment plant – a miniature version of the dilapidated set-up I'd seen in Jajmau, this one newer, and clearly working. 'This is our chrome recovery facility,' Asad says. 'We recover the chrome, and most of the treated water we reuse for washing, cleaning.' We stop at a tank of clear, stale-looking water. 'This is our effluent. If we are throwing this into the river there is no problem!' Asad has recently gone one further, he explains, trialling the use of anaerobic digesters to help process the plant's other waste. We find them in another part of the factory grounds: two bulging plastic sacks, like distended stomachs, connected to a series of gas pipes. Inside, bacteria feed off the waste liquid to create biogas, which can be used to power the factory. 'This pilot is successful, so we have very safe disposal of the waste, and green energy.'

Back in his boardroom, Asad is visibly frustrated. Even

the responsible tanneries, he says, are being tarred with the same brush as a handful of rogue operators. 'The government should be supporting us. We have invested a lot of money, and yet they closed our production. The bureaucrats said, "Tanneries? No!"'

On the drive back from Unnao, I yell for our driver to stop. On the roadside, the puddles are luminous chartreuse green. Chromium. A man sitting on a wall nearby chewing gutka explains that someone had dumped chromium sludge here. The city eventually removed it, but now the puddles glow whenever it rains. Stories like this are common in Kanpur. The chief suspects are poorer tanneries, looking to avoid having to pay the treatment plant for waste processing. In the nearby villages of Khan Chandpur and Rania, where unscrupulous factories have dumped an estimated 45,000 tonnes of chromium sludge, groundwater samples have found chromium at 4,000 times the WHO safe limit for drinking water.[48]

A little way further, we stop the car again. Inside the entry to an open lot are mounds of what from a distance look like heaps of black mud. In fact, on closer inspection, the mounds are blue, and still steaming: chromium sludge. Toxic waste, dumped on the roadside. The landowner, an elderly man with a thin moustache who gives his name only as Mr Salim, comes over. Salim is an animal feed manufacturer – he dries leather scraps to be ground into meal. I ask about the dumping. 'From the factories,' he says, with an air of resignation. 'To avoid the authorities they come in the night time and dump it here.' It is not the first time they have dumped sludge on his land. The culprits, he says, are waste haulers, 'labourers who get the contract to remove all the waste material and take it to the [municipal] landfill, but once they get the money they just dump the waste anywhere.' Salim has reported the incident to the police, but so far nobody has come. We stare for a while at the blue mass of toxic sludge. Even if they do come, who will remediate the soil beneath? How many toxic dumpsites like this exist in Kanpur and Unnao alone?

The true level of chromium-6 pollution in the Ganges is difficult to measure. Numerous studies have detected heavy metals in the Ganges around Kanpur that far exceed safe limits for human and ecological health.[49] In 2015, Indian researchers found that the Ganges at Kanpur exceeded recommended safety limits for lead, cadmium and chromium, the latter at nearly fifty times India's legal limit set.[50] The problem is not limited to Kanpur – according to one study, chromium-6 exposure nationwide costs Indians more years of healthy life expectancy than multiple sclerosis, Parkinson's disease, and several types of cancer. But determining the number of illnesses, or deaths – or even exactly how much chromium is being emitted into the Ganges, and by whom – is still frustratingly unclear. The infrastructure around Kanpur, as in the rest of India, is still creaking and outdated, despite the Namami Gange project's efforts to purify the river.

A team of scientists at the Indian Institute of Technology Kanpur are working on one possible solution: real-time monitoring of pollutants in the Ganges. One morning in Kanpur, we drive over to the university's department of mechanical engineering, situated within a wide and leafy campus on the edge of town. The place is still deserted due to the pandemic, but I find Professor Bishakh Bhattacharya and his colleagues in their offices, tinkering with electronics. Bishakh – a warm and good-tempered man in an ill-fitting checked jacket, baggy chinos, and sneakers – was inspired to work on Ganges pollution following his own father's death. 'I was doing the last rites of my father in Varanasi,' he explains. 'The water was so filthy that I was thinking: how can anybody do this in the river?'

Under a canopy in the department's garden, Bishakh and a younger engineer, Satya Prakash, show me what they are working on: a large bright yellow buoy, designed to measure the pollutants in the river in real time. 'We have sensors installed in it that collect seven parameters – data like oxygen dissolved in the water, pH, conductivity,' Bishakh says. The buoy collects both data and physical samples, 'so that we can know every single day the current condition of the river'. This system is just a prototype for now, and unable to test for specific chemicals – instead using water conductivity as a shorthand for metal content. Even then it is already proving useful. During the pandemic, for example, the team saw in real time the effect that lockdowns were having on pollution in the river: with tanneries closed, dissolved metal content fell by more than 60 per cent.[51] 'It was good, very clean,' Satya says. 'And now the industries are working again, all the pollutants have started getting back into the system.'

The remote system is needed, Bishakh explains, because existing data collected by the state pollution control boards tends to be poor quality, and often inaccurate. 'There is no record of exactly how much industrial waste is being dumped, how many carcinogenic elements are being put into the water. The cities like Kanpur get a filtered drinking water supply. But the villages keep taking water from the river, and people are getting sick.'

Bishakh and his colleagues hope that by deploying similar sensors along the length of the river, they might be able to build a kind of real-time pulse monitoring system for the Ganges, able to detect major pollution events as they're happening. 'We can see: how pure is the water before the city and after the city? Where are the hotspots? Where is most of the industrial waste being dumped?' There are still challenges to overcome. The team hopes to add more detailed sensors, including those for E.coli, to better measure pathogen pollution. There's the issue of funding. And there's the security issue: during trials, a student had to watch the station at all times, lest it be harvested for scrap.

Bishakh is supportive of the Namami Gange project, and has already seen the benefits. 'The river quality has improved. There are fewer floating pollutants – the temple flowers, the dead bodies,' he says. But without a robust detection method, he fears that the shutdowns in Kanpur will just move the polluting industries downstream. He shakes his head. 'These people do not follow any rules.'

What if our ecosystems can't clean themselves, as we once thought? Then the idea of threshold theory collapses. In truth, we've long understood this reality. As early as 1912, residents of the Toyama prefecture in Japan began reporting severe pains in their joints and spines. The mysterious sickness, which became known as *Itai-itai* disease, was cadmium poisoning, caused by a nearby mine releasing waste into rivers. The toxic metal accumulated in the river's fish and plant life, and in the villagers' rice. *Itai-itai* is now remembered as one of the four 'Big Pollution Diseases' that blighted Japan in the early twentieth century, all caused by industrial waste.[52] The Japanese disease outbreaks are sometimes used to illustrate an alternative to threshold theory, known as boomerang theory. As in: *what goes around, comes around.*

This is where we now find ourselves. In some cases, such as sewage in the Ganges and other rivers, the sheer quantity of pollution is causing tipping points to be passed. In others, the assumption that nature could metabolise all our waste is being found wanting. Plastics, for example, do not so much biodegrade as come apart over the course of decades, in the process filling the environment with micro- and eventually nanoplastic particles. Cadmium, lead, and mercury bioaccumulate in soils and sediments where they can be taken up by microscopic organisms such as algae; these are in turn eaten by fish, eventually passing back up the food chain into humans.

The same is true for the entire class of chemical toxicants known as Persistent Organic Pollutants (POPs). These chemicals – which include PCBs, brominated flame retardants, and legacy pesticides such as DDT, which Rachel Carson wrote about in *Silent Spring* – are poorly soluble in water but highly soluble in fats, making them prone to absorption in animals. Twenty-eight classes of POPs are now banned under the United Nations' Stockholm Convention, a treaty signed by 179 countries.[53] POPs have since been found on Mount Everest and in the Arctic, and are still regularly found in the oceans and the tissues of living things even decades after being banned. As recently as 2012, ninety-four birds in St Louis, Michigan, were found to have died from DDT poisoning – forty years after the pesticide was banned in the USA.[54] In 2017, researchers from Newcastle University found 'extraordinary' levels of PCBs in the body of amphipods (a bit like shrimp) inside deep ocean trenches;[55] it is currently theorised that the chemicals are slowly passed downwards by microorganisms in the water column before accumulating in deeper waters. One place that PCBs are notorious for accumulating is in whale blubber;[56] in Canada, Beluga whale carcasses have been found so contaminated they are classified as toxic waste.[57]

We are still discovering exactly how durable some substances can be. Take, for example, per- and polyfluoroalkyl substances (or PFASs). A massive class of industrial chemical used for their non-stick properties, PFASs are found in an astonishingly broad array of household items: non-stick pans, raincoats, stain-resistant fabrics, pizza boxes, paint, cosmetics.[58] Their unwillingness to break down and tendency to bioaccumulate in the environment has earned them the nickname 'forever chemicals'. Exposure to PFASs is linked with cancer, liver damage, asthma, and thyroid dysfunction, among a whole host of health complications. In studies, PFASs have been found in the bloodstream of 99 per cent of people, and even inside unborn children.[59]

PFASs have been in use since at least the 1940s. In the 1950s, the chemical giant DuPont began using one such compound, the surfactant perfluorooctanoic acid (PFOA) in the manufacture of Teflon. Despite knowing for decades that PFOA caused liver enlargement and cancer in lab animals,[60] DuPont pumped thousands of litres of PFOA-laced wastewater into rivers in Ohio, and into unlined 'digestion ponds' on its premises, which contaminated the drinking water of more than 100,000 people.[61] The true extent of DuPont's deception did not really emerge until 1999, when the lawyer Robert Bilott filed a suit against DuPont in West Virginia on behalf of a farmer whose cattle were dying after consuming creek water loaded with PFOA leaking from a hazardous waste dump. To date, DuPont and Chemours, its spin-off company (allegedly set up to help reduce its legal exposure from PFA pollution), have settled lawsuits against them worth over $753 million. (In 2009, DuPont replaced PFOA with something called GenX, which – wouldn't you know – has also been linked to cancers and is now banned by the EPA.)[62]

PFASs operate through a particularly insidious mechanism, by tricking the body's endocrine system. Endocrine-disrupting chemicals (EDCs) – which also include BPA, dioxins, and phthalates, a class of chemicals ubiquitous in plastics – mimic or inhibit hormones, even passing into cells, and affecting DNA.[63] BPA, for example, interacts with oestrogen receptors. As a result, they are particularly dangerous for pregnant women and for children. Exposure to EDCs has been linked with increased risks of miscarriage, fertility problems, and cancer.[64] Children exposed to EDCs in the womb or at an early age show increased risks of ADHD, obesity, diabetes, and early puberty. Yet EDCs are everywhere: in our food packaging, plastics, toys, cosmetics, cleaning products, in the very walls of our homes. There is growing evidence that the rise of these chemicals in everyday life may be linked to a widespread rise in endocrine-related disorders, such as obesity. (In 2006, the endocrinologist Bruce Blumberg coined the term to describe such chemicals: 'obesogens'.) There is also evidence that being exposed to EDCs during pregnancy and early childhood can disrupt boys' sexual development. In Italy, where PFOA and a similar chemical, PFOS, have contaminated drinking water for thousands of people, exposure to the chemicals in young men has been linked to shorter penises, lower sperm counts, and lower sperm mobility.[65] Pthalates, a type of EDC commonly found in plastics, have been linked to falling sperm counts worldwide.[66]

Endocrine-disrupting chemicals like PFASs confound the threshold theory of pollution,[67] in that their effects do not require large exposures – conversely, they can harm even at infinitesimal doses. In 2009, the US EPA set an advisory limit for PFOA and PFOS, two of the best-known forever chemicals, of 400 parts per trillion (ppt). In 2016, it revised those down to 70 ppt. In 2022, as a result of growing evidence of toxicity, the agency revised those down again sharply – *17,500-fold* in the case of PFOA, to 0.004ppt.[68] And yet, the EPA admits that PFASs are currently found in the tap water of more than 200 million Americans,[69] and are widely found in drinking water around the world. Because they are so ubiquitous in manufactured products, PFASs accumulate in waste. They are found concentrated in sewage sludge and in landfills, in compost, and in emissions from incinerators, to the extent that exactly how to dispose of them has set off a minor crisis in the waste industry.[70] We are starting to realise that some things can't just be washed away.

On my last morning in Kanpur, I visit another tannery. From the outset, it's clear that the building has seen better days: the lights in the stairwell don't work, and there are exposed wires poking through dusty brickwork. The owner, Kaushlesh Dixit, is an older man with whitening hair and weathered features. We're joined by another tanner, Mohammad Arshi, a younger man in a thick insulated coat. With all parties still worried about the coronavirus, we sit on plastic chairs in the car park and talk. Kaushlesh wraps his scarf around his face to fight the cold. I soon realise I'm shivering. After a few minutes, someone emerges with paper cups of hot, milky chai from a nearby street vendor, which we use to warm our fingers. 'The leather industry is dying in Kanpur,' Kaushlesh says. He has worked as a tanner for three decades, and has seen the government restrictions against the leather trade slowly increase. The pandemic, the fifteen-day on-off mandate, and now the religious festival closures, he says, are making it impossible for most tanneries to stay in business. 'There is not a problem of quality, or price – we are still getting demand consistently from our buyers. The problem is since we have reduced the quantity of our production, the big buyers are turning to other sellers.'

Kaushlesh, like many other tannery owners I've spoken to, refutes that the problem with Ganges pollution is solely down to the leather trade. He points out that there are other heavy industries in Kanpur, including metalworks and electroplating, which both use chrome. 'The authorities have been continuously blaming the industry leaders, and saying that tanneries are the root cause of the pond polluting the River Ganges,' he says. 'We have been continuously asking: please show us where our wastewater influence is going to the Ganges. Tell me: "This company is discharging their waste water." We will set fire to that company!'

Mohammad, visibly angry, blames the government. 'They hunt the Muslim tannery owners,' he says. As with the waste pickers in Delhi, most tannery owners in Kanpur are Muslim, and have been subjected to increasing persecution since Modi's party took office in 2014. 'This government is promoting Hindutva [Hindu nationalism]. Ninety per cent of the tannery owners are Muslim, and the Ganga is the Hindu sacred river. So they hunt us, and blame us, saying, "You are polluting."'

'It's all politics,' Kaushlesh says. 'All politics.'

I tell Kaushlesh about the chromium dump in Unnao, and about the luminous green puddles. Even then, he insists tanneries are unlikely to be the perpetrators. 'The BCS [Basic Chromium Sulphate] factories outside of the city are doing it,' he says. He blames corruption within the utility companies and pollution board officials in charge of the CETP. 'The authorities have given it to a private contractor to run, and they are cutting down the treatment days and saving a lot of money,' he says. 'If I am discharging all my wastewater to the CETP but the CETP is discharging that wastewater to the Ganges, how can you blame me?'

Even though I don't agree with them – plenty of evidence

points to the tanneries as the main source of the chromium in the Ganges, and anyway, it is their existence here that has drawn the sulphate factories – it's hard not to sympathise. We as consumers are outraged at the stories of industrial pollution, and yet it's our hunt for ever-lower prices that pushes the most polluting industries to cities like Kanpur. Since the pandemic, Kaushlesh has seen dozens of tanneries shutting up shop and moving away. Instead, they are moving their businesses down-river to West Bengal, and in some cases as far as Bangladesh, where environmental regulations are even more lax. Rather than cleaning the river, or preventing chemical pollution, Mohammad says, the crackdown is simply moving it.

Therein lies one of the major challenges facing the Namami Gange project. In my three weeks in India, I saw plenty of evidence of Modi's cleanup campaign. In Varanasi, the Hindu holy city, I toured gleaming new sewage plants, and saw the government-funded efforts to remove plastic and rotting flowers from the river. In Kanpur, the government has capped several municipal sewage drains that once emptied into the river, instead sending it for treatment. But it was also more than evident that the waste flowing into the river is, so far, outpacing anything being built to stop it. The challenge is as much spiritual as it is physical. Restoring the Ganges to anything like its normal state requires more than money – a change in attitudes, in a collective will.

It's getting late in Kanpur, and I have a train to Delhi to catch, so we say our goodbyes to Kaushlesh and Mohammad. In the cab on the way back to the hotel, my colleague Rahul spots something, and so we pull into a lot on an otherwise ordinary street behind a low-slung row of market stalls. It's a waste dump, several football fields in size. The entire place is carpeted with leather scraps, the colour of a pale horse. The dried-out hides bend and crackle underfoot, water pooling in cracks between the skins, the same vivid green as on the roadside. Goats and chickens are picking through the waste for food. It's a desolate sight, like something out of a Cormac McCarthy novel. A monument to slaughter. I wander among the dead skins, thinking about how little we truly see of the way things are made, and how little we understand of the true cost.

For the first time in my entire journey, I feel sick.

CONTROL, DELETE

'Every machine, one way or the other, will die'

In the lobby of Fresno airport is a forest of plastic trees. *A bit on the nose*, I think, as I pass through arrivals. This is central California, home of the Grand Sequoia national park, with some of the oldest and most iconic trees in the world. But you can't put a 3,000-year-old redwood in a planter (not to mention the ceiling clearance issue), and so someone at the tourist board has clearly deemed it fit to build these towering and impressively convincing fakes. I pull out my phone and take a picture, amused and somewhat appalled. What will live longer, I can't help but wonder, as I press my hands against their petroleum bark: the real trees, or the fakes? Did anyone ever ask who will recycle these artificial giants when they're done? Or will they, like the forest itself, eventually burn?

I haven't come to Fresno to see the trees; I've come about the device on which I took the picture. In a blank-box warehouse among a cluster of blank-box warehouses in the south of the city, bright green trucks are unloading pallets of old electronics through the bay doors of Electronics Recyclers International, Inc (ERI), the largest electronics recycling company in America.

Waste Electrical and Electronic Equipment (better known by its extremely unfortunate acronym, WEEE) is the fastest growing waste stream in the world. Electronic waste amounted to 53.6 million tonnes in 2019, a figure growing at about 2 per cent a year.[1] Consider: in 2021, tech companies sold an estimated 1.43 billion smartphones,[2] 341 million computers,[3] 210 million televisions, and 550 million pairs of headphones.[4] And that's ignoring the hundreds of millions of appliances, games consoles, sex toys, electric scooters and sundry other battery-powered devices we purchase every year. Most of our electronic devices are not disposed of, but live on in perpetuity, tucked away forgotten in drawers, cupboards, and lofts. My kitchen drawer at home has at least two old iPhones and three pairs of headphones in it, which we have kept 'just in case'. In this, we are not unusual. As the head of Music Magpie, a UK smartphone refurbishing service, told me, 'Our biggest competitor is apathy.'

Globally, only 17.4 per cent of electronic waste is recycled. As for the fate of the remaining 82.6 per cent, it probably won't surprise you to learn by now that we simply don't know. Between 7 and 20 per cent is estimated to be exported, and around 8 per cent thrown into landfills and incinerators in the Global North. The rest is unaccounted for. And yet, WEEE (no, I can't do it) is, by weight, among the most precious waste there is. One piece of electronic equipment can contain as many as sixty elements – not only commodities like iron, copper, aluminium, but a host of rare earth metals including cobalt, neodymium, and tantalum, which are used in everything from motherboards to gyroscopic sensors. A typical iPhone, for example, contains around 0.018g of gold,[5] 0.34g of silver, 0.015g of palladium, and a tiny fraction of platinum.[6] That might seem small, but in metals terms, you'll find ten to fifty times more copper in a tonne of electronics than in a tonne of copper ore,[7] and 100 times more gold per tonne of smartphones than ore from even the most productive mine.[8] Multiply by the sheer quantity of devices, and the impact is vast: a single e-waste recycler in China, GEM, produces more cobalt than the country's mines each year.[9] By one estimate, up to 7 per cent of the world's gold reserves may currently be contained in e-waste.[10] Research by United Nations University has estimated that the value of materials in our e-waste is worth approximately £50.9 billion a year.[11] As a result, several large mining and metals companies, including Glencore and Umicore, have now set up e-waste recycling divisions, recast as 'urban mining'.

Aaron Blum, the co-founder and chief operating officer of ERI, arrives wearing the unmistakable corporate uniform of a tech executive: a navy hoodie and jeans. 'You're going to need these,' he says, handing me a bright orange pair of earplugs. Aaron's a handsome guy with a soul patch, young-looking for an executive type. He and a friend started up ERI back in 2002,[12] when he was just out of college. 'My last semester I had an internship for a computer resale company,' he says. 'My job was to go around San Diego and buy up material that was coming off refresh [being upgraded] for banks, healthcare, those types of things, refurbish it and resell it.' At the time, California had just passed a series of laws banning electronics from landfills due to hazardous chemical contents. But little electronics recycling infrastructure existed. 'I didn't know anything about electronics. I was a business major,' Aaron says. Today, ERI has eight facilities across the US, and processes 57,000 tonnes of scrap electronics every year. 'We service every zip code, every industry you can think of.'

The factory itself is a cavernous place, spread over more than 11,000 square metres, the roof held up by bright yellow pillars. 'On a nationwide basis, we process around 20 million pounds [9,071 tonnes] a month,' Aaron says. Around 250 ERI employees work at this facility, dismantling and shredding old tech to get at the valuable materials inside.

To get onto the floor, we have to pass through a security scanner. I hold up my notepad and dictaphone, but am waved through. Security here is tight for a reason: millions of dollars of still-functioning or repairable electronics pass through this warehouse in a year, making it a tempting target for thieves. In the loading bay, a goateed guy named Julio is unloading pallets of shrink-wrapped monitors from a Salvation Army truck. Charity shops are a major source of ERI's product – anything the charity shops can't sell they bale up in their central warehouses and eventually send here. 'We have the whole country with regard to Salvation Army,' Aaron says. 'We probably have twenty-five Goodwills. We've got all the big box retailers. We've got over seventy OEMs [Original Equipment Manufacturers], a lot of banking. We're just getting into the healthcare side.'

The factory workers unload the pallets and scan them, before passing the contents off to be dismantled and sorted. 'You can't shred certain materials, so you've got to do a sort,' Aaron says. Electronics are piled all over the place: flat screens, desktops, printers, keyboards, DVD players. At a set of tables, nine men are taking apart large flat-screen TVs, their electric screwdrivers emitting a low *whizzz*. Another is, less delicately, smashing a monitor apart from its casing with a hammer. ('Due to the adhesive,' he says.) The dismantling crews, Aaron says, will typically dismantle up to 6,500 pounds [2,948kg] of devices in a day. 'The hardest thing is a flat screen,' Aaron says. 'But commodity wise, there's probably double, maybe even triple what I'm able to recover off a CRT.'*

We pass a noticeboard marked FOCUS MATERIAL, upon which actual parts have been pinned as visual aids: whole mother boards, wire scraps, memory cards, monitor casings, different grades of wire. 'We've learned over time this hits home more often than reading a document,' Aaron says, reeling off the contents as we pass.

Scrap recycling contains so many different materials that, to simplify things, the industry has over time developed its own shorthand vocabulary. Light copper, for example, is 'Dream', No. 1 Copper Wire is 'Barley', while insulated aluminium wire is the delightfully onomatopoeic 'Twang'.[13] There's no such poetry here, however. Instead, the extracted pieces are thrown into boxes scrawled with things like COPPER, SERVER BOARDS – HIGH GRADE and CAT 5 WIRING. Inside one box I notice a coil of LED Christmas lights. 'During the holidays we get a ton of these. This is all copper, in the wire,' Aaron says, grabbing a handful. 'We have to go through and manually cut the bulbs off.'

Some materials – paper, batteries – must be removed for safety reasons. 'If something gets through that can't be shredded, you can have a fire, or an explosion,' Aaron explains. 'When you're shredding metal, it gets really hot.' Heat-sensing cameras constantly scan the factory floor for hot pockets. I notice the workers are all in masks and gloves. E-waste also contains any number of toxicants, ranging from lead and mercury to polybrominated flame-retardants and PFASs. 'We still do get PCBs, even though they're not manufactured any more.' Aaron stops at a bin, and pulls out a long thin lamp

* A cathode-ray tube, used in old TVs and monitors.

taken from inside an LCD monitor. 'The goal is to remove these mercury tubes. Then we'll send it downstream to a lightbulb recycler.'

The piles get larger further into the factory. I spot a huge heap of old computer screen housings, desktop computer housings, chunks of various white goods. Judging by their boxy shapes, most are at least a decade old; throwbacks to a time when every piece of electronics had to be black, white or grey. The pile looks sketched in charcoal, or as if a bomb went off in a PC World circa 2005.

The centrepiece of the facility – the *coup de grâce* – is the shredder. A hulking beast of a machine, it stretches along the length of the building, three storeys high, making a prodigious racket. (Hence the earplugs.) Once the waste has been sorted, a worker in a Bobcat telehandler carries it over to the conveyor's gaping maw, to be obliterated. Inside the machine, ultra-hardened spinning blades cut through aluminium and plastic like ice in a blender. 'When you're shredding electronics, you're creating dust that does contain lead from the circuit boards, so we have collection hoods sucking up all the dust,' Aaron hollers. The dust has to be disposed of as hazardous waste. I nod, exhilarated by the sheer violence of it.

Along the length of the shredder, a series of magnetic belts, air-sorters, and filters automatically size and separate the materials as they pass, dropping each through its own chute and into giant 'super sacks'. We stop at one sack and look down at a treasure haul of silver-grey flecks. 'We call this precious metal fines,' Aaron says. 'This is gold, silver, and palladium. It comes from the circuit boards.' This single sack's contents are likely worth enough to buy a decent-sized car.

Further along the line the conveyor splits off into tributaries. A robot arm whirs above one, picking up parts. 'We used to have about fifteen pickers on this line. Now you've got one managing the robot, and two or three pickers,' Aaron says. The company spent a lot of money training the robot, which picks far faster than any human could, and is now 97 per cent accurate. Aaron seems to prefer it to people. 'It comes to work every day, and it never got Covid,' he says, as if he's joking, but I sense possibly not. Towards the end of the line, more metals are rolling off into their own supersacks. ERI's biggest material streams are steel and plastic, by weight, followed by aluminium and brass. The circuit boards are sent to LS Nikko, a metals manufacturing giant based in South Korea; the aluminium goes to the US smelting giant Alcoa. 'The steel might go to your large steel buyers in the US – they might send it off to their mills in Turkey or whatever it might be. But otherwise, everything stays domestic.'

Once upon a time, ERI's entire business relied on selling the metals they recover on the international commodity markets. However, after a series of market crashes threatened to put the business under, the company changed course. 'You can't play that game,' Aaron says, 'you're going to run yourself out of business.' Today, ERI charges its customers a fee for disposal, dismantling, data removal, and recycling. Most of its customers are motivated not by reducing waste, he says, but by cybersecurity. 'Ninety-nine per cent of the electronics you have today have your data on it, from your wearables to your iPads, to your tablets to the TV hanging in your living room. So data has become very, very important,' Aaron insists. Paranoid about losing industrial secrets to China, companies would rather have their old machines wiped and shredded, a service ERI is more than happy to supply for a fee. 'We have Homeland Security come to a couple of our facilities. They will escort the material to the shredder, stand watching the shredders while we run the material through, and sometimes they'll even take the shred out.'

We're passing back through the factory when something catches my eye: an entire pallet of TV screens from a major manufacturer, still neatly boxed and plastic-wrapped. They appear to be brand new, because they are. Even so, they're here to be shredded. 'Chances are this is probably from the manufacturer,' Aaron explains. 'They don't want this product resold and competing against their new products, so they want it all destroyed.'

I'd expected to see this at ERI, but perhaps not so brazenly. Manufacturers and retailers routinely destroy returns and unsold items, known as deadstock, en masse. As Kyle Wiens, the founder of the global electronics repair chain iFixit, told me, these 'must-shred' contracts are the 'dirty secret' of the recycling industry. ('The recyclers are desperate for manufacturer contracts, so they'll do anything and keep their mouth shut,' Wiens said.) In 2021, for example, an investigation by ITV News in the UK found that Amazon was sending millions of new and returned items per year to be destroyed.[14] Just one of the company's twenty-four fulfilment centres was destroying up to 200,000 items a week, ranging from TVs to books to Apple Airpods. (Amazon says it has since stopped the practice.) In 2020, Apple sued a Canadian recycler for reselling some of the 500,000 devices it had sent for shredding.[15] The recycler, GEEP, blamed rogue employees – but the implication that the devices had been in good enough working order to sell rightly set off a wider scandal.

The unfortunate truth is that companies destroy new and nearly new products all the time. Luxury brands and technology companies, propelled by a constant churn of new products, are reluctant to heavily discount or donate unsold items that might undermine the sales of new models. The fashion brand Burberry, for example, admitted to incinerating £18 million pounds a year of unsold items between 2014 and 2018, to prevent them from being sold at discounted rates.[16] In other cases, the financial upside of processing unsold items or returns is not worth the costs (setting up processing centres, paying wages, repairing faulty items), so it's cheaper to write it off. Burn it or bury it, wasting is cheap.

There's an old axiom, commonly heard in both the recycling trade and retirement homes, that *they don't make things like they used to*. This is partly good old-fashioned nostalgia, but it's also something real, intuited in our everyday lives and easily proven in any junkyard or landfill. Goods cheaply bought are cheaply made – no surprise there. But when it comes to e-waste, that sense is often followed by another, more serious allegation: that of 'planned obsolescence', by which industries design products with artificially short lives, so that they need to be replaced more quickly.

Planned obsolescence is often treated as some kind of conspiracy, when in reality it is just a historical fact. The idea first gained traction as early as the 1920s, when manufacturers were coming up with means to convince people to upgrade to newer models. Christine Frederick, an American household economist and popular author behind books like 1929's *Selling Mrs Consumer*, defined 'progressive obsolescence' as 'a readiness to "scrap" or lay aside an article before its natural life of usefulness is completed, in order to make way for the new and better thing.'[17] Rather than something negative, planned obsolescence was pitched as a progressive force; during the Great Depression, obsolescence was touted as a way to restart the flagging economy. One early advocate, a real estate broker by the name of Bernard London, even argued for the creation of a government agency that would prescribe *maximum* (rather than minimum) life spans for products, at which point they would have to be traded in for a new model.[18] (Reading it now, London's plan looks remarkably similar to many tech trade-in schemes.) Obsolescence went radically against the existing notions of the time, in which businesses had competed to build products that were of the highest quality; Henry Ford, for example, built the Model T to be repairable and long-lasting, while declaring, 'We want the man who buys one of our cars never to have to buy another.'[19]

But obsolescence, like disposability, really experienced its golden age in the mid-1950s, as the post-war boom began to slow. Manufacturers were realising that once everyone had a TV, refrigerator, car, and radio, they needed to create reasons for them to 'upgrade' to newer models. With durable goods, that required making goods, well, less durable. As Victor Lebow, a marketing consultant writing in the *Journal of Retailing*, put it in 1955: 'Our enormously productive economy demands that we make consumption our way of life, that we convert the buying and use of goods into rituals, that we seek our spiritual satisfactions, our ego satisfactions, in consumption . . . We need things consumed, burned up, worn out, replaced, and discarded at an ever increasing rate.'[20] Waste was no longer just a side-effect – it was the goal.

Not everyone signed on. One of the most vocal critics of this new credo was the American journalist Vance Packard, who had gained fame for his 1957 bestseller *The Hidden Persuaders*, an early exposé of the advertising industry. In his 1960 follow-up, *The Waste Makers*, Packard railed against businesses that had pushed disposability and obsolescence, creating the waste crisis of the era. According to Packard, obsolescence comes in three forms. There is 'obsolescence of desirability', when products go out of fashion for aesthetic reasons; 'obsolescence of function', in which a product is surpassed by newer technology (think of smartphones going from cellular to 5G in less than twenty years). Finally there is 'obsolescence of quality', when products are designed to wear out faster or within specified time limits. Perhaps the most infamous example of the latter is the so-called 'light bulb cartel', when in 1925 representatives from the world's largest light bulb manufacturers – including General Electric, Osram, and Philips – met in Geneva to discuss the problem of slumping sales. Up to that point, light bulbs had been so long-lasting as to rarely need replacing (one carbon filament bulb that hangs inside the fire department in Livermore, California, has been burning almost non-stop since 1901).[21] So in Switzerland, the light bulb companies agreed to collectively reduce the designed lifespan of light bulbs from around 2,500 hours to just 1,000 hours. It was the first of a series of quality downgrades over the following decades. As Giles Slade writes in *Made to Break*, General Electric estimated that by reducing the lifespan of flashlight bulbs from three battery lives to one, 'it would result in increasing our flashlight business approximately 60 percent.[22] Making products worse, counterintuitively, was good for business.

By the time Packard was writing *The Waste Makers*, creating products with ever-shorter lifespans was a widely accepted way of creating consumer demand. As Harley Earl, then head of design at General Motors, put it in 1955, '[I]n 1934 the average car ownership span was 5 years: now it is 2 years. When it is one year, we will have a perfect score.'[23]

Today, overtly planned obsolescence of the kind practised by the light bulb cartel would be seen as unethical, not to mention illegal. But the fundamental tenets of obsolescence that Packard railed against are ingrained in the practices of modern industrial design. We now recognise that 'obsolescence of desirability' is simply another way of describing fashion, and is created not only by the advertising industry that Packard so despised, but through collective culture. (Fast fashion would have blown Harley Earl's 1950s mind.) 'Obsolescence of quality' would now be called *design lifespan*, the number of uses that appliances are manufactured to withstand. Companies spend millions building elaborate labs to test products for durability – artificial feet that simulate the action of shoes pounding pavement, or rigs that run tumble-dryers for thousands of hours until the mechanism wears out. Durability influences price; we understand that when we buy a cheap knock-off on Amazon it's not going to last as long as something that costs three times as much. We also understand that some obsolescence is good: replacing cars for models with more fuel-efficient engines or catalytic converters, for example, or plastic bottles that contain fewer toxic chemicals. Similarly, we know the rapid churn of smart devices in the last decade-plus has largely been driven not by faulty products, but by the relentless pace of technological progress. (When Arthur C. Clarke wrote in 1962 that 'any sufficiently advanced technology is indistinguishable from magic' he might have been describing the iPhone.)

Even so, the electronics industry in recent years has faced allegations that a modern version of planned obsolescence is contributing to our rising tide of e-waste. The assertion is less one of fact than of feeling, but nonetheless has slowly but steadily accumulated, along with our rubbish. In 2017, for example, Apple admitted that it had been quietly using software to slow the performance of older iPhones. The company claimed the measure was to protect the devices from electrical issues associated with ageing lithium-ion batteries;[24] critics argued that by hiding the problem, Apple had likely induced people to upgrade their phones early. This assertion was somewhat backed up by the fact that when Apple discounted battery replacement fees that year by way of apology, iPhone upgrade rates fell sharply.[25] Apple, which has faced multiple lawsuits as a result of 'Batterygate' – including a $500 million civil action lawsuit that it settled in 2020 – eventually apologised.[26] But it has also engaged in a pattern of behaviour that critics allege undermine its self-image as a sustainable business. In 2017, for example, Apple sued a small independent Norwegian repair shop for replacing broken iPhone screens with refurbished ones imported from China,[27] claiming they were counterfeit. The iPhone 13, introduced in 2021, initially included a feature that would disable the phone's Face ID unlock system when the screen was replaced with one not made by Apple itself.

This is a small part of a wider trend. As our electronic appliances have become more internally complex, we have become ever more divorced from how they function. Inwardly and outwardly, they have become black boxes. Most of us would have no idea how to fix our own phone when it breaks. This is not only of our doing: over time, many of the largest manufacturers have removed the ability for ordinary consumers to replace even commonplace parts such as batteries. Apple attempts to prevent you from even opening its devices, through the use of proprietary screws. Instead, tech companies argue that repairs must be conducted by trained professionals, and in many cases by the company itself – for a hefty fee, of course.*

---

* I use Apple here not for any past grievance, but because it is the world's richest and best-known tech company; many of its rivals have employed similar tactics.

iPhone owners who want to repair their own phone at home, for example, must now pay a $1,210 deposit to hire Apple's special tools.[28] I find this development particularly disheartening, because as a teenager in the mid-2000s I spent my weekends working at a mobile phone repair stall in the local shopping centre, happily swapping out dud batteries and broken screens from old Nokias and Motorolas for new ones. In those days, repairing phones and even certain laptop components was easy: one could pop out a Nokia 3210 battery and replace it in seconds, injecting the phone with a new lease of life.

But it isn't just amateurs that find modern electronics hard to repair. In the race to make our devices ever thinner and cheaper, our smart devices have become fundamentally more difficult to fix: once-removable parts printed onto circuit boards; screens held in place by adhesives; tiny earbuds that cannot be opened at all. Then there are software locks, where older devices no longer receive software or security updates, rendering them unusable; Android phones, for example, are typically only supported for two years. Hannah and I still own an iPad Mini that is perfectly functional, but which cannot use most of its apps due to outdated software support. Thus, it has been relegated to the forgotten pile.

This fight over repair has come to a head in recent years, thanks largely to the campaigning of organisations like iFixit (which in addition to having an international chain of repair shops, publishes detailed manuals and How To guides online for free), the Restart Project, and Right To Repair Europe. In France, new electronics must now be labelled with a 'Repairability Index' score, which rates products on categories like access to manuals, spare parts, and ease of access; at the time of writing, the European Union is planning to roll out a similar measure continent-wide.[29] In the US, a number of states have begun to introduce Right to Repair laws, mandating that manufacturers make manuals and spare parts available to independent repairers and the public after release – a rule that already exists in some jurisdictions for vehicles and white goods.

The fight goes far beyond smartphones and tablets. In the US, farmers have sued the tractor maker John Deere for allegedly using software and other means to prevent them from repairing and maintaining their own agricultural machinery.[30] Tesla, which uses software to 'lock' its cars' battery range and performance unless customers pay extra, has taken action against owners using software to remove these limits; the customers, understandably, responded by questioning at that point whether they own the car at all.[31]

These fights can sometimes feel obscure, or ridiculous; most of us are probably not going to attempt to fix our phones, even with a $1,200 repair kit. But the fight over repair has real-world consequences further afield – often in places where a Genius bar is much harder to find.

Rich countries have been exporting e-waste to poorer countries almost as long as there has been e-waste to send. Exporters from the US and Europe began shipping used electronics to China along with scrap metals and plastics as early as the 1980s. But the trade didn't earn much attention until 2002, when the Basel Action Network released *Exporting Harm*, a now-infamous documentary about the environmental crisis that e-waste was inflicting upon recycling towns in southern China, particularly the town of Guiyu. The film showed desperately poor informal workers, including children, breaking down electronics by hand, burning the casings off wires, and separating components with crude acid baths, in order to access the valuable scrap metals inside. The ecological and human toll of the e-waste trade in Guiyu was heartbreaking. Soil and water samples in the recycling zones contained lead and other heavy metals that exceeded every WHO threshold; in one study, 81.8 per cent of children under the age of six surveyed were found to be suffering from lead poisoning.[32]

The Chinese government has since cleared many of the informal recycling shops in Guiyu, and concentrated e-waste inside allocated industrial zones.[33] But while China's imports of e-waste have fallen, the amount of e-waste we produce has only grown. For the last few years, the most notorious destination for Western electronics has not been China but Agbogbloshie, a slum in Accra. Dubbed 'The World's Largest E-waste Dump' by a number of publications, Agbogbloshie has been the subject of harrowing coverage in *The Guardian*, *Washington Post*, *Al Jazeera* and *Vice*, as well as the subject of numerous viral YouTube films (most of them shot by white Westerners). I remember being horrified by the images when I first saw them: barefoot 'burner boys' torching scrap wire, the toxic fumes billowing from scorched earth, others cracking open imported mobile phones against the backdrop of a dilapidated slum. Once again, it seemed, Western waste electronics were being dumped on the world's poor, who were reaping the toxic consequences. So I decided that I needed to see it for myself. Once again, it turns out that the reality is not quite so simple.

It's a glorious day in Accra when I arrive outside Evans Quaye's electronics shop. 'Welcome!' Evans, who is expecting me, steps out to offer a warm handshake. A spectacled man with a bright smile and taste for even brighter shirts (today's number: floral, greens and reds, worn with chinos), Evans is an electronics importer. He makes his living buying used laptop computers from Holland to resell in Accra's thriving second-hand market.

'Our biggest market is schools,' Evans says, gesturing into the shop, an open-fronted unit with sun-baked brickwork and faded signage, on the end of a row of similar shops. Inside, I spot a pile of several dozen new-looking Dell boxes, stacked chest high. Children have recently returned to classrooms after the pandemic, and orders are picking up again. 'Some of these have come from schools in Holland, and now they will go to schools in Ghana. Come,' Evans says, gesturing at the high sun, and perhaps noticing the sweat quickly pooling in the nape of my neck. 'We'll talk in my office.'

Evans's office is a few blocks away, so I climb into the passenger seat of his Volvo. I notice more repair shops as we drive. Outside one, rows of old Sony televisions hide in the shade of an awning. Outside another, kitchen appliances spill into the street: washing machines, refrigerators, irons, almost all of them imported. 'You will see them repairing TVs, printers, microwaves, irons – anything you can think of,' Evans says. Ghana's economy, like many in this part of Africa, is built on the second-hand trade. Every year, more than 1.2 million containers pass through the nearby port of Tema, loaded with second-hand goods from the US, Europe and Asia. Not only electronics, but clothing and cars, too.[34] In 2009, the last year with solid data, Ghana imported 215,000 tonnes of electronics, 70 per cent of it used.[35] The imports are by necessity, as much as anything: the minimum wage in Ghana is just 12.53 Cedis (£1.28) an hour; so like many countries in this part of West Africa, few people can afford to buy new.[36] That's where repairers like Evans come in.

We arrive at Evans's office, a small one-room building opposite a school. It's a cool, welcoming place, the desk dotted with old laptops, a ceiling fan looping lazily overhead. One of Evans's employees arrives bearing bottles of cold water, and I drink thirstily. Evans has worked in the second-hand trade since he left school, in 2002. These days, he is a rep for SNEW BV, a 'circular telecoms' company based in Holland, which collects used electronics from across Europe for resale. The newer models are resold in Europe, the older ones in Africa, where prices are lower. 'The standard model we receive is five years old. But in my experience, we can use a machine for as much as fifteen years. I have a Pentium IV here . . .' Evans pulls out an old Dell laptop, which must be at least a decade old (Intel stopped making the Pentium IV processor in 2008). 'I've been using it a very long time, and it's working perfectly.'

The Basel Convention, an international treaty signed in 1992 to prevent the transit of hazardous waste, prevents the international shipment of broken electronics, which might be classified as garbage. So every device that Evans brings into Ghana must first be checked in Europe. 'It's killing my business, because in America they can send machines without cost,' he says. (The US has never ratified the Basel Convention.) 'So I don't compete with the machines from the US.' For Evans, being unable to move broken machinery is a problem, because broken devices can still be useful as a source of spare parts. 'We have a lot of good repairers in Ghana,' he says. 'Our problem is a lack of parts.'

The following morning Evans drives me across town to Danke IT Systems, a small repair shop on the second storey of a strip mall. It's a tiny place, internet café-style, with a handful of machines set up for customers. A sign on the wall reads: JESUS IS LORD. The manager, a bright eyed and bald 39-year-old named Wisdom Amoo, is seated behind his desk with a laptop on his lap and a screwdriver in his hand. The cubby holes and drawers around him are brimful of laptops and parts: Dells, mostly, but also machines from HP, Lenovo, Asus, Apple.

Wisdom has just finished the HP in his hands, which had a broken charging port. The part is soldered down, so he has jury-rigged a solution by converting a display port to accept a charging cable. 'I need to cut a hole, here, and replace it with parts from another machine,' he says, gesturing with a precise finger. Certain computer models, he explains, tend to have the same issues: screen burn in one, faulty trackpads in others. Wisdom gestures to a rack of shelves at the back of the store. 'These ones are mostly broken,' Wisdom says. They're next in the queue.

Repairing is as much an art as a science. Organisations like iFixit often provide instructions and how-to videos online, but those resources cannot always be accessed in Ghana. As a result, Ghana's repairers are a tight-knit community of often self-taught engineers. 'Sometimes you have to use your brain, too. Sometimes there might be a problem you've seen before, sometimes you go to YouTube and have a little idea,' Wisdom says. Repair work is a delicate skill; a single slip with a soldering iron can ruin a laptop, rather than fix it. When he's soldering, Wisdom holds his breath.

On the rare occasion a machine can't be repaired, he'll pay a small amount to the customer and harvest it for parts. The rest will be picked up by the recyclers. 'One machine we can use to fix a lot of different machines,' Wisdom explains. But over time, the task is becoming more difficult. In recent years, laptops and tablets have started to ship with processors and memory that are welded to the motherboard, rather than as separate pieces. To make the machines thinner and cheaper, the companies now fix parts in place with adhesives or solder, where once would have been screws. 'The new models, you can only upgrade the RAM. Some of the parts, let's say the USB port, replacing them is very, very difficult.' For a 5-year-old

Dell, Wisdom explains, a new screen might cost 450 Cedis (around £45). But for newer models with bonded or touch screens, the entire display has to be replaced. 'This one: 2020 MacBook Air, 3,500 Cedis,' Wisdom says – around £350. (iFixit, which rates new electronics on repairability, has scored a number of Apple products 1 out of 10.) 'Some touch screens, like [on a] Microsoft Surface, even opening the model can cause it not to work.' With iPads he won't even bother.

That's a worry, because the alternative to buying used electronics in Ghana is buying new, cheaply made products from the brightly lit, stuff-filled Chinese superstores that have started to appear across the city. He recalls how recently, a customer had bought a cheap Chinese laptop. ('For an i3 processor, 2200 Cedis. I told him: I sell i5s for 1,500.') Three months later, the man appeared in Wisdom's shop. The machine had broken. The superstore blamed the damage on a power surge – an everyday reality in parts of Ghana – and refused to replace it under warranty.

'There are no replacement parts for those,' Evans says. 'So where do you think it ends up? The dumping site. So now who pollutes the environment: me, offering a refurbished machine that is going to last five, ten or fifteen years, or the new one, which will break down in less than a year?' In Accra, he explains, the scrap recyclers from dumps like Agbogbloshie are part of the repair ecosystem. 'If the repair shops had a machine that cannot be repaired, then the scrap boys would come around to pick up those machines and take them to Agbogbloshie,' Evans says. 'Then the repair shops, when they needed parts, would normally go down to Agbogbloshie to see if they could source parts from there. If I need a part for a TV with a working screen but a broken power system, by chance, I might find the same TV with a broken screen but the power system working. So we can harvest the parts, and repair the machines.' Only after usable parts had been extracted would the remainder be dismantled and sold off for scrap.

This, Evans explains, is the context often overlooked in Western media stories about Agbogbloshie. E-waste is not coming to Ghana to be dumped. Rather, it's coming to be used. The dumping narrative is further undermined by research from the United Nations Environment Programme, which found in 2011 that up to 85 per cent of e-waste in West Africa was being created by domestic usage of 'new or used' electronics, not foreign imports.[37]

In that sense Agbogbloshie was not, despite widespread reports to the contrary, 'the world's largest e-waste dumpsite'. It was not even a dumpsite, although the wider slum contained a landfill, as most do, without formal waste collection. Rather, Agbogbloshie is a neighbourhood, home to schools, markets, churches. It is also home to a large informal settlement, Old Fadama, which is home to an estimated 100,000 people, many of them immigrants from the poor northern regions of Ghana. The 'dump' was a scrapyard – albeit a very large and well documented one, and one where the environmental controls were, as they are in many poor, informal settlements in the Global South, tragically lacking.

I'm writing in the past tense because Agbogbloshie[38] no longer exists – at least, not in the form that it once did. In 2021, in the midst of the Covid pandemic, the Ghanaian police raided and demolished the scrapyard. A couple of days after my meeting with Evans, I head there to see it for myself. From Old Fadama, I can look out across the Odaw river to where Agbogbloshie – the dump, but also the homes around it – once stood. The site has been razed. Bare earth covers where the former scrapyard and shops once stood, a handful of heavy earth movers still dragging topsoil around. The government supposedly plans to use the land to build a hospital.[39]

I don't intend to minimise the pollution caused by e-waste at Agbogbloshie, which was nothing short of horrifying. The toxic toll of burning and dismantling the e-waste on the dump polluted the soil, the groundwater, the workers, and even the food. In 2011, a Ghanaian researcher found that soil at a nearby school exceeded European safety standards twelvefold;[40] in another study, eggs from chickens living in the settlement contained 220 times the tolerable daily intake dose of dioxins.[41] The informal recyclers, who work without even the most basic safety protections, were repeatedly found to have unsafe levels of lead and cadmium in their blood and urine. Agbogbloshie might not have been the largest e-waste dump in the world, or even close to it, but it was almost certainly among the most polluted.

With Agbogbloshie gone, many of the scrappers have simply crossed over the river into the informal settlement known as Old Fadama. Fadama itself is a sprawling place: colourful wooden dwellings separated by thin mud lanes, so close as to be almost on top of one another. Inside, some inhabitants sleep eight to a room. Few of the buildings are plumbed with toilets or running water. The scrap workers have set up shop around the edge of the slum, on the beach. There we find several dozen men (and they are all men) at work dismantling waste: hammering apart old engine blocks and tearing down refrigerators. Here: a boy who can't be more than a teenager, cutting up a gearbox. There: an older man, greying, prising the springs from an old car seat. With nowhere to keep it, the scrappers store their stocks in the open. The evidence of the dismantling is everywhere. There are piles and piles of kitchen appliances. One tangle of old bicycles looks like the aftermath of a collision at the Tour de France. The ground is flecked with snapped fragments of television casings and chunks of old motherboards, which chickens and goats pick through, looking for lunch.

The burner boys have set up as far from the houses as possible, out on the river beach beyond where a crowd of children are playing football. A dozen of them are gathered around a makeshift fire pit, carrying nests of wire on metal poles, which they press into the flame. The plastic melts away like marshmallow, giving off smoke. The air is singed with the wretched stench of plastics and burning solder. I want to talk to some of them, but my colleagues advise me not to. Since the government clearance, some of the scrap workers have become angry with Western interlopers, whom they justifiably blame for the government's decision to knock down their old homes. 'They have given thousands of interviews,' Evans says. 'They were still evicted.'

But Evans has known many of the scrap boys for years, and offers to introduce me to some at his office. I expect him to mean one or two. Instead, when I turn up at SNEW the next day, he has gathered more than half a dozen young men – some of whom I'd still consider kids – to speak to me. They file into the room, looking down, wearing flip-flops and the tattered shirts of rich European football teams: Juventus, Chelsea, Real Madrid. Most of them are not from Accra. 'We're all from the north,' Yakubu Sumani, a wiry young man in tired black jeans and a brown T-shirt says. Yakubu is one of the oldest and, it immediately becomes clear, the spokesperson for the group – largely because many of them don't speak English, instead speaking Dagbani, a language widely spoken in Ghana's north.*

---

* Ghana has somewhere between fifty and eighty indigenous languages, though only a few are spoken widely.

Yakubu explains that he had worked in the scrapyard since he was fifteen, earning around 15-20 Cedis (around £2) a day, buying and selling material. It wasn't easy or glamorous. But it paid better than other jobs in the informal sector; many of the young men were able to earn enough to send some money back to their families in the north. 'You break, you buy, you sell,' he says. Companies – 'Big men' – would come by once a week to buy material in bulk. 'You don't know what the person you sell it to will use it for,' he says.

Yakubu recalls the day that the police arrived to clear Agbogbloshie. 'They came with weapons. They were just arresting us. They beat some of us,' he says. Ever since, the scrappers have scattered across the city. Some returned home; there are scrap jobs in the north, too. 'They don't want us to be there. They want us to be away,' he says – meaning, out of Accra, for good. 'We are scattering around. Everyplace you go, the police chase us.'

'We have a lot of people who are displaced,' Evans says, quietly.

By destroying Agbogbloshie, the Ghanaian government has not eliminated the e-waste, but spread it across other informal settlements throughout the city, and further out. 'The waste is still in the system. But where is it now? You can't find it, because it is scattered all over.' Evans and many of the scrap traders have tried to argue that it would be better to formalise the scrap trade in Ghana, as China and others have done: allocate a designated industrial zone, provide health and safety rules, and give the workers formal recognition and social support, such as pensions. 'None of them have any savings,' he says. 'What they make, they eat that night.' Instead, he fears that the country will soon follow in the footsteps of others, including China, India, Thailand, and Uganda, and ban the import of used electronics entirely. 'That is my fear here in Ghana now. Because if it happens here, we are doomed.'

Any e-waste import ban, Evans says, would not stop the e-waste ending up in Ghana's landfills. It's not going to stop people from buying new phones and laptops, and discarding the old. It would, however, likely destroy the second-hand sector, meaning the loss of thousands of decently paid jobs. 'Recycling, one way or the other, creates some kind of pollution. But processing raw material is even more environmentally unfriendly,' he says. There are 1.2 billion people in Africa. As their economies grow, they will rightly begin to buy and waste new stuff at similar rates to consumers in the Global North. What will be the environmental cost of their e-waste, in countries with little to no recycling infrastructure? 'At the end of the day,' Evans says, 'if I'm able to give a machine a second life, we're going to save the machines from going to Agbogbloshie or somewhere else to pollute the environment.'

Too often, the way that we talk about e-waste falls into a kind of guilt trap: *aren't we terrible, for inflicting our waste on others*. But the story is rarely that simple. To simplify waste exports as 'dumping' erases the local people doing the importing of waste, and ignores the reasons why they might want it in the first place. Too often our tendency towards simplistic saviour/demon narratives blur a subtler and harder truth: that waste exports are a complex system with deep ties to colonialism, poverty, and trade. That isn't to say we should permit dumping, but rather recognise that for consumers in the Global North, our role in this story is more difficult. (And that we aren't always the protagonist.) It's also recognising that the decisions we make about how we design and dispose of products have global impacts, many of which we rarely see. A more serious attitude to e-waste might ask why Extended Producer

Responsibility schemes – in which technology companies pay into a central fund that goes towards recycling and product end-of-life programmes – aren't sending far more money into the Global South, where their devices end up. When we discuss the Right to Repair and the obsolescence of products, we rarely see the last links in the chain, the people who often use those products the longest. Who is listening to their voices? Where are they at the table? As Adam Minter has written, 'When you think about it, insisting that Africa's second-hand traders adopt Europe's definition of "waste" . . . is a kind of colonialism.'[42]

As I step out of Evans's office into the bright sunlight, I'm reminded of something he'd said that first morning we'd met. 'Every machine one way or the other will die,' he said. Then he grinned that irresistible grin. 'Like humans: everything has a lifespan.'

THE DAM BREAKS

'When you take away all of the environmental and
safety regulations, this is what you get'

On 25 January 2019, the miners at the Córrego do Feijão iron ore mine outside Brumadinho, Brazil, were taking their lunch break in the cafeteria when they heard a loud and sudden crack.[1] Some of the miners surely assumed it was nothing. One later told reporters he thought a tyre had burst. But others must have known at once what the sound meant. A few hundred metres uphill from the cafeteria, the dam was breaking.

I find it hard not to imagine the miners' terror in that moment. The dam was holding back an estimated 11,700,000m³ of waste rock and ore left over from the mining process, known as tailings, mixed with vast quantities of water into a thick and toxic slurry. The dam itself was actually built out of settled tailings, and thus was little more than compacted dirt built up over decades into a series of stepped mounds. The miners knew that such 'upstream' dams, although common in Brazil, are vulnerable to liquefaction: when a sudden shift in load or vibration can transform the sludge from a near-solid to a liquid form. When that happens, there's nothing you can do.

At 12:28 p.m. the dam wall breached. There exists video

footage of this moment, which is both horrifying and tragic to watch. The dam face, one moment a steep and grassy bank, suddenly collapsed into a massive wave of rust-coloured liquid. Trees at the base of the dam were flattened like grass underfoot. A tsunami of sludge flushed downhill at speeds up to 80km/h; in places, the crest towered 30m high. For the miners there would have been a roar, a brief chaos, and then: nothing. The cafeteria and administrative buildings were washed away within seconds. Some workers outside in their vehicles tried to flee, in vain. The tailings razed the mine and its railroad, and rushed downhill, destroying homes, bridges, and farmland, eventually flooding into the Paraopeba river. Two hundred and seventy-two people, including almost all of those workers in the cafeteria, were killed.

In its wake, the tailings left behind a ruined Martian landscape. As it passed, the wave deposited a thick layer of metal-rich sludge, stretching over 8km from the mine to the river. The tailings that flushed into the Paraopeba killed off its fish[2] and left the water unsafe to drink.[3] A later federal investigation into the collapse found that the rupture had been triggered by a build-up of unseasonal rainfall, but was ultimately caused by unsafe dam management by Vale, the mining giant that operated Córrego do Feijão. In 2020, Brazilian prosecutors charged sixteen people, including Vale's former CEO, with murder and environmental crimes, alleging that it 'systematically concealed' the dam's safety risks. In 2021, the company was ordered to pay $7 billion in compensation.[4]

Tailings dams are an essential feature of mining, although one most people rarely hear about until they fail.[5] Depending on the mine, tailings – the unprocessed ore, rock and dirt left over from the extraction process – can contain a variety of toxic pollutants, and even radioactive materials. Rather than be ejected into the surrounding area, where they might pollute the environment, they are most often interred inside massive impoundments, supposedly permanently. Nobody knows exactly how many tailings dams exist worldwide, although I have seen estimates as large as 21,000.[6] The Global Tailings Portal, an initiative set up by the Church of England's pension fund as a response to Brumadinho, has gathered data on 1,862 tailings dams operated by some of the world's largest mining companies. Of those, 687 were classified as 'high risk'; 166 have had previous stability issues.[7] (This is likely to be an underestimate, as it relied on voluntary disclosures by the mining companies, and included virtually no data from China, which is host to more mines than any other country.) And, while most dams are safe, failures happen with alarming regularity. Between 2009 and 2019, eighteen major dam accidents occurred in twelve countries, according to Brazilian researchers. There are eighty-seven dams of the 'upstream' design that failed at Brumadinho in Brazil alone.[8] The country's mine regulator has called them 'ticking time bombs.'[9]

As mining facilities tend to be remote, the number of deaths in dam collapses tend to be small. The environmental damage does not. In 2015 a similar dam collapse at another iron ore mine in Brazil, in Mariana, killed nineteen people and unleashed 43,700,000m³ of mine tailings, in what has been called the country's biggest ever environmental disaster.[10] In 2000, a gold mine near Baia Mare, Romania, spilled an estimated 100,000m³ of cyanide-containing wastewater into the Danube river basin, killing off masses of wildlife and contaminating the water supply of over 2.5 million people.[11]

Mining waste is not something many of us probably think about too often in the Global North. Today, the raw materials we rely on are extracted thousands of miles away, in countries like China, Australia, Brazil, and Peru. In the UK, most of the country's mines closed decades ago, and are now a distant memory. And yet you likely create mining waste almost every moment of every day. It is just one of many unseen forms of waste that we enable indirectly, through consumption. Consider: copper ore typically might contain just 0.6 per cent copper,[12] meaning that you need to mine a tonne of ore to produce a kilogram of copper – a metal which wires our houses, is found in most of our electronics, and upon which modern life depends. That yield is even smaller for rarer metals: platinum ore, for example, can yield as little as 2 grams of platinum per tonne. Mining enough gold to manufacture a single wedding ring is thought to generate 18 tonnes of waste.[13] Then there are the wastes formed during the extraction process: gold, for example, is often separated from surrounding rock with cyanide, a potent poison that is known to leach into groundwater and pollute rivers. Cyanide-laced tailings, impounded into dams, can breach and cause large-scale disasters. Many smaller artisanal mines[14] use liquid mercury to extract gold from the surrounding earth; gold mining is the largest contributor to mercury pollution worldwide.[15] All told, there are thought to be 3,500 large-scale mines worldwide, producing 100 billion tonnes of solid waste per year.[16]

Mining waste has reshaped our landscapes in profound and visible ways. Confined tailings ponds become new lakes. Solid mine waste – spoils and slag left over from extraction – is simply piled into great heaps. In Britain, these mounds can go by many names: spoil heap, slag heap, bing, batch. Drive through any former mining region in the world and you will see them: conspicuous mountains, as if freshly burst from the earth. In Loos-en-Gohelle, France: five heaps, the tallest 180m high. In Heringen, Germany: 'Monte Kali', a 250m-high white mountain of salt. In Catalonia, Spain: a 500m tall pile of salt and potash waste that has grown taller than the 'natural' mountains that surround it. In South Wales, where I went to university, there are so many slag heaps from the now abandoned coal mining industry – around 2,000, in all – that at first I struggled to distinguish them from the mountains. These constructs are often toxic, and can be dangerous. Wales, for example, lives with the memory of the Aberfan disaster in 1966, when a coal tip on a mountainside collapsed, crushing a primary school, killing 144 people, 116 of them children.

Sometimes these man-made structures, these dumps, take on new meanings. The heaps in Loos-en-Gohelle, for example, are now a UNESCO world heritage site. In West Lothian, just outside Edinburgh, five towering shale oil waste bings known locally as the Five Sisters have become a haven for rare flowers and mosses.[17] In Hirschau, Bavaria, you can go skiing on a mountain of leftover sand from a Kaolin mine. Looking at these strange monuments, I couldn't help but think back to Vic and Jamie's landfill. Again and again, we are scouring out the innards of the Earth, and filling the empty void with our waste.

I'm driving south on Route 69, on the trail of mining waste, when the satnav pings to inform me that I'm not in Kansas any more. I've come to the Tri-State district, a patch of the Midwest where the northern tip of Oklahoma cuts in greedily between the southern border of Kansas and the eastern edge of Missouri. Mine country. It is also tornado country, and when I pull up at the headquarters of the Quapaw Tribe Environmental Office, in the tiny town of Quapaw, Oklahoma, the last breaths of the previous day's storm are still shaking the trees. I've come to Quapaw to see about a ghost town – or more precisely, ghost *towns*. You see, while the Tri-State district used to be mine country, it is now better known as the home of the Tar Creek Superfund site, one of the largest and longest-running environmental disasters in America.

Summer King, an environmental scientist with the Quapaw tribe, meets me at the front desk. 'Tar Creek is my baby,' she explains. An intensely likeable woman with freckles and short dyed-red hair, Summer wears the downy jacket and sturdy boots of a scientist who spends most of her time in the field, rather than behind a desk. This land has belonged to the Quapaw Nation since 1838, after the American government forced them from their previous home on the Arkansas river. (The name Quapaw means 'downstream people'.) Summer is not Quapaw, but Cherokee. 'We have eight of us in the office. Three Quapaw tribal members, two of us that are Cherokee, and the rest of them are non-native,' Summer says, leading me into a cosy meeting room decorated with maps and mining photographs.

In 1848, prospectors struck ore across the state line from Quapaw in Joplin, Missouri. They discovered that the underlying rock, a band of chert and limestone laid down around 320 million years ago called the Boone formation, was rich in both lead and zinc. Mining began. In 1912, prospectors struck an even richer seam here in Oklahoma, triggering a lead rush. Entire mining towns sprang up as people flooded in to get at the newfound riches: Picher, Cardin, Century, Douthat, Zincville. By the mid-1920s, the Tri-State district was producing more lead and zinc than anywhere in the world.[18] According to local lore, every American bullet fired during the World Wars was mined here in Oklahoma; death as primary export.[19]

At their peak, the mines in the Tri-State area were cavernous things, 'roomy enough to accommodate a fleet of greyhound buses', as a journalist for *Gems and Minerals* magazine wrote in 1968.[20] The miners here used both wet and dry methods. The exhumed rock was crushed and separated, to remove the metals. The fine tailings left over were poured into vast tailings ponds; the dry waste fragments of rock and gravel, called 'chat', were dumped on the surface in piles hundreds of feet high. These chat hills came to dominate the area for miles. Photographs from the time show a desolate, alien terrain, with entire towns built among the pale and crumbling massifs. By the 1970s, the mines had all but given out, and the few remaining mining companies abandoned the area, leaving the chat piles – still containing toxic lead, zinc and cadmium – on the surface. When the pumps that drained the mines were finally turned off, the mines flooded. The water reacted with newly exposed sulphuric compounds in the rock to produce acid mine discharge that killed off marine life and stained the stream, Tar Creek, red.

Beginning in the 1970s, evidence began to build of the toxic impact of the lead remaining in the chat piles. 'Most of it is limestone gravel, but of course there's still lead, zinc and cadmium in it,' Summer says. As the years passed, teachers in the schools in and around Picher noticed that many of the children lagged far behind the state average on test scores, and struggled with their attention spans – a classic sign of lead poisoning. 'In those days, in Picher, the houses went right up to the bases of the piles. It's where the kids went to play,' Summer tells me. 'People filled their sandboxes with it, so now it's in the yard. You swam in the ponds around the piles, which are full of dissolved metals. You'd get people coming through on the piles with their dirt bikes.' She shakes her head. 'So it was just continuous exposure, for their whole lives.' Lead accumulates in teeth and bone, and leaks into the brain, the liver, and the kidneys. At high levels, lead poisoning can cause developmental problems and cognitive defects, increase risk of miscarriage, and cancer. The lead levels at Tar Creek were among the highest in America.

In the mid-1990s, soil samples taken in Tar Creek found elevated lead levels in schools, daycare centres, and parks. Sixty-five per cent were above the recommended safety limit of 500mg/kg for lead; the highest sample contained twenty-one times the safe dose.[21] Several also exceeded the safety limits for cadmium. Heavy metals were also found contaminating several drinking wells, poisoning the water. In 1993, researchers at the University of Oklahoma took blood tests of 189 Native American children from the Tar Creek area. Fifty-three per cent had blood lead levels above the dose deemed 'safe' at the time, and some were more than three times what was then the limit. (In 2012, the US Centers for Disease Control halved the blood lead reference value for children from 10 µg/dL to 5µg/dL. In 2021, it lowered it again, to 3.5 µg/dL.[22] According to the EPA, 'No safe blood lead level exists' and 'even low levels cause harm'.)[23] The poisoning was not limited to the children. In adults, residents of Tar Creek have been found to have elevated risks for lung cancer, tuberculosis, kidney disease, hypertension and a range of lung problems. One operating theory is that lead and other metals were entering their bodies in dust from the chat piles, carried on the wind.

In 1980, in response to a number of high-profile pollution scandals – including the infamous Love Canal incident, when a toxic chemical dump was found to be poisoning a school in upstate New York – the US Congress passed the Comprehensive Environmental Response, Compensation and Liability Act, better known as Superfund. The new law provided federal resources for environmental cleanups. In 1983, shortly after its passing, the Environmental Protection Agency created a 'Superfund National Priorities list' of the most toxic sites in America which required urgent remediation. Tar Creek was on the list. It still is. 'We're getting ready to celebrate our fortieth birthday on the Priorities List next year,' Summer says.

In Quapaw, Summer takes out a geological map of the area and spreads it out over the dark wood table. 'Everything in red is where they mined,' she says. The map is startling: so much of the Tri-State district has been hollowed out that it looks like the earth itself has a disease.[24] Black marks indicate the presence of known mine shafts; there are hundreds of them, pocking the paper like lice on a scalp. When lead was first discovered here, not all of the Quapaw, who owned the mineral rights, agreed to lease their land to the miners. 'So the mining companies would go to the Bureau of Indian Affairs, and the BIA would have that tribal member declared incompetent and sign the lease on their behalf,' Summer says. 'That's the *original* environmental injustice out here.'

We clamber into Summer's truck and drive west onto the Superfund site. Tar Creek is a massive area – 1,188 square miles, encompassing two creeks and several former mining towns, now all-but abandoned. The roads are straight, deserted, and lined with trees: cottonwood, black willow, sycamore, red cedar. 'Everything out here is invasive,' Summer says. 'It's the only thing that survives the metals.' We haven't gone far when we first spot the chat: great pale hills rising from the plains. At the top of one mound, a man in an excavator is digging at the loose rock, causing gravel to cascade down the slope. In places, the finer rock has bonded with rainwater into strange outcrops, like sandcastles mid-collapse.

The chat piles themselves all have numbers, but the bigger ones have names: Pioneer, Slim Jim, St Joe, Bird Dog. Most are named after former mine shafts, or the respective landowners.

Others, Summer isn't sure. 'You can call one pile about five different things,' Summer says. 'It drives my EPA project manager absolutely insane.' At one point, there were more than 300 chat piles across the Tar Creek area, containing an estimated 165 million tonnes of waste, though in truth nobody really knows.[25] There was simply too much to count.

When the mining companies left, the task of remediating Tar Creek fell to the Quapaw. Since Tar Creek was named a Superfund site in 1983, work has been underway to remove the chat. Most of the chat is first sold to gravel processors, for recycling. 'There's one approved use for chat, and that's asphalt,' Summer says. 'They really like the chat because it's very hard. On the Mohs scale [of hardness] it's a nine, so one of the only things harder is diamonds. When you put it in asphalt, it doesn't wear.' The roads out here are paved with chat; the older buildings were built on chat, mixed into the foundations. 'We called it chatcrete. It's amazingly hard. It takes a guy with a jackhammer to break it up,' Summer says. The chat processors remove any marketable materials, and dump the remaining rock, which the Quapaw are in the slow process of collecting, into a central repository.

We drive over to the repository, a wide chat hill with a viewpoint over the local terrain. 'In the mining days, they processed here. This was called the Central Mill,' Summer says. There is still evidence of the old mines here and there: a wooden wheel, what looks like a trace of a rail track. The repository itself is built atop a tailings pond 50ft deep; waste all the way down. 'EPA looked at it and said that it was so big, so deep, that it would break Superfund to clean it up. So they did some engineering and decided that rather than cleaning it up, we could put 10 million tons of chat on top of it and create a repository.' At the moment, the Quapaw crews are hauling up to 6,000 tons of chat from the piles to be consolidated here.

We get out of the truck at the summit. I crouch to pick up a small triangular piece of chat. It's hard and cold between my fingers. Uncomfortable. 'You can see how angular it is, how hard. That's why the processors love it – it makes great roads,' Summer says. When it's piled up, the larger pieces of chat are pushed to the surface, like the nuts in a bowl of granola (this phenomenon, granular convection, is sometimes called 'the Brazil nut effect'). 'And then the fine stuff all accumulates in the middle,' Summer explains. 'That's where the metals are, so it's the most hazardous.' We look out at the piles. Trucks and loaders scramble up distant slopes, loading and unloading, slowly dismantling the horizon. The scale is breathtaking. From up here, the pale puddles around the chat look like globs of dropped paint. I think of T. S. Eliot's *The Waste Land*: 'What are the roots that clutch, what branches grow / Out of this stony rubbish?' What forces exist great enough to shape such a landscape? Volcanoes, tectonic plates, glaciers, rivers – and people.

'I like to say that Tar Creek is the poster child for unregulated industry,' Summer says, tossing a piece of chat back to the ground. 'When you take away all of the environmental and safety regulations, this is what you get.'

It is at this point in the book that I have to admit to what is, if not a deception, then at least an omission. (If this were a thriller, here comes the twist.) At the beginning of this book, I wrote that every British person throws away 1.3kg of waste per day. This is not a lie, per se: it is the best estimate provided by the World Bank, a relatively estimable institution, and based on solid data. But it is also not quite the truth, as it relies upon two misdirections. The first is the subject: *every British person*.

This, of course, is not literal, but an estimated per capita average – it divides the total amount of waste discarded by the total population and again by 365, to get a nice round number. This is a common habit in environmental policymaking; it makes things personal. This tactic can be used to foster engagement (or, as we saw in the case of recycling, deflect blame). But it also creates an illusion that we are all equally wasteful, when in fact waste is deeply unequal. Just as the private jet-owning ultra-rich produce far more carbon emissions than the ordinary person, not everyone produces the same amount of waste.

The second deception is in the noun: *waste*. As we know, waste is a matter of perspective. But in this case, waste specifically means 'household waste', i.e. the things we throw away at home. It does *not* include industrial waste, or all the waste that went into the manufacture and transport of the product to get it to your home in the first place. Industrial waste – the offcuts and castaways from factories, smelters, farms, chemical plants, and the sundry other acts of extraction and human enterprise required to make our stuff – is the great secret of the waste industry. It is the elephant never in the room. Industrial waste, unlike our household waste, is not collected in rubbish trucks and shipped off to be weighed at MRFs. Instead, it is amassed on private land, in tailings ponds, private landfills, effluent ponds, and chat piles. It is flushed into our rivers and expelled into our air. Very little data exists on it: what it is, where it is, how toxic it might be. The United States Environmental Protection Agency (EPA), for example, breaks down household waste for every person in the USA every year, but does not publish regular data on industrial waste. What's more, policymakers dispute what even constitutes industrial waste, and therefore what should be counted. The EPA, for example, defines 'non-hazardous industrial wastes', but *excludes* the waste leftover from oil and gas production, coal ash, and all mining waste, despite these amounting to the largest source of waste, by weight, than any other (and, as in Tar Creek, much of it hazardous).*

Exactly how much industrial waste is there, globally? The truth is: we don't know. One regularly cited estimate states that 97 per cent of all waste is produced by industry, not households. That figure, quoted regularly by waste campaigners, is based on an estimate made by the US EPA in 1987.[26] But the figure is outdated, vague, and almost impossible to verify. More recent estimates that reflect modern recycling practices and data collection show the gap is smaller (although not by much) and depends heavily on how industrialised a country is. Canada, a country with extensive mining and forestry sectors, produces just 35.5 million tonnes of household waste per year, but 1.12 *billion* tonnes of industrial waste.[27] (In pure coincidence, this fits almost precisely with the 97–3 estimate.)[28] In the UK, a relatively deindustrialised nation, 'commercial and industrial' wastes amounted to 43.9 million tonnes in 2018, plus a further 57.5 million tonnes of 'non-hazardous construction and demolition wastes'. Household wastes, by contrast, totalled just 26.4 million tonnes – that is, just 11.8 per cent of the total.[29]

This is the reality about waste, that must be reckoned with: for all our focus on household recycling rates, for all the effort we spend washing out yogurt pots and collecting bottles, the vast majority of waste happens upstream, before our products ever get to us. The writer Joel Makower calls this the 'Gross National Trash' – the waste built into everything we buy, and which dwarfs that which we throw away. In addition to food

---

* These are collected under a separate category, 'Special Wastes'.

waste, for example, the global agriculture industry generates 4.5 billion tonnes of animal faeces (shit), most of which is stored in surface effluent ponds, the run off from which is choking our river systems.[30] And yet we rarely see that waste, even though it is embedded in the food we eat.

I remember feeling defeated the first time I learned about this problem, like the air was sucked out of the room. Faced with the scale of industrial waste, individual actions can begin to feel pointless, a grain of sand in a chat pile. That is not my intention in including it here. Nor is it to place the entirety of our waste problem at the feet of industry (though they deserve more responsibility than they willingly take). Rather, it is simply to acknowledge the inherent waste within everything we make and consume, so often thoughtlessly, so that we can at least start to have honest conversations about waste, and the true scale of our wastefulness. The waste in that new smartphone is not just in the cardboard box, or the phone itself. It is in the land defiled to extract the rare metals, in the forests cut down and discarded to make the packaging, in the toxic chemicals discharged into rivers to manufacture the plastics inside. It is in the people and the places that we lay waste *to*.

Rebecca Jim moved to Tar Creek in 1978, or 'the year before Tar Creek changed colour', as she puts it. Rebecca is a slight woman, softly spoken, with greying hair pulled back and her eyes neatly framed by dark tortoiseshell glasses. We're sitting in her office at the LEAD Agency, the environmental charity that she founded in 1997 to campaign about lead poisoning in Tar Creek. It's a cosy place: a white clapboard house in Miami (pronounced Mi-am-ah), a small town a short drive outside Quapaw. The porch is filled with arts and crafts made by local children, every shelf and cranny crammed full of books on environmental law and records of the Superfund site. Rebecca, who is Cherokee, first moved to the Tri-State district to take a job as a Native counsellor with the local schools. 'I loved being here. It was invigorating. But I found it challenging with some children. We had a lot of children that had trouble concentrating, or sitting still, or learning,' she says, her voice soft. 'Later, we found out it was probably lead poisoning that was causing our children to have such difficulty.'

Although the EPA declared Tar Creek a Superfund site in 1983, it wasn't until the 1990s that regulators started taking the realities of lead poisoning here seriously, largely due to work by the local Indian Health Service, who had begun testing children's blood levels. Even then, it was years before any serious action was taken. 'The largest lead processor here also had a lead paint company. So instead of getting a turkey at Christmas, the workers got paint. So that was also exposure, in their homes,' Rebecca says. When the levels of lead poisoning became clear, there was confusion about exactly what the source was. 'The mining companies wanted to prove that it was the lead paint and not chat that was poisoning the children, to lessen their liability.' (Studies showed that dust from chat and soil was the most likely source, although lead paint was likely an additional way that the poison took hold.)

In 1995, the EPA began to do yard remediation in and around the Tri-State mining towns – digging up gardens found to have dangerous levels of lead, and replacing them with clean soil.[31] At the same time, community groups educated local families about the dangers of chat. The chat, meanwhile, stayed in place. In 2004, in part down to continued political pressure from campaigners like Rebecca, the state of Oklahoma offered home buyouts to families with young children. In 2006, the state extended those buyout offers to the rest of the families living in four former mining towns – Picker, Cardin, Zincville, and Hockerville – after a technical study showed that the towns' foundations had been extensively undermined, and were at high risk of subsidence and sinkholes.[32]

The buyout process itself, however, was bitterly fought. That year a category 4 hurricane tore through Picher, killing six people, and destroying many of the remaining houses. In response, the company in charge of the buyouts lowered their offers, giving residents the choice: claim their insurance money, or accept the buyout. Two hundred and sixty-two residents sued. 'There were a number of people that didn't want to leave,' Rebecca says. 'Once the offers started coming it, they were not fairly done. It divided people – those that got enough money to leave severed close ties with other people, who didn't get as much as they deserved. It caused a lot of anger towards the government, but also anger between each other.' Those that did receive buyout money were largely white families, who were leasing the land from their Quapaw owners. 'They didn't get paid for the land, they just got paid for the house,' Rebecca says. 'The Indian landowners didn't get any money for that, just the homeowner or renter. This story is never told. So they were left with a house-less piece of property that they can't use. They got nothing.'

This story is all too common. All over the world, indigenous people are often the ones who bear the brunt of our industrial waste – from the destruction of rainforest in places like Brumadinho, to the ravaging of tribal lands across North America by mining companies. In the US, for example, several tribal nations are still grappling with the aftermath of the government's uranium mining during the Second World War, which blighted tribal lands and left a legacy of injustice.[33]

Picher and Cardin are no longer, officially, places. By 2009, almost all the residents had abandoned their homes, and the state of Oklahoma disincorporated the towns. Public functions stopped. Only a couple of elderly holdouts remain.

One afternoon, Summer and I drive into Picher, or what remains of it. Most of the buildings have been demolished. 'They bulldozed the old houses and dumped them in a sinkhole,' Summer says. Squares of chatcrete now mark where they stood, water mains protruding from cracked foundations. In the centre of Picher is a memorial: a statue of a gorilla – the local school's mascot – surrounded by placards telling the story of the mines in an oddly triumphal tone. The school is one of a few buildings still intact; law enforcement now use it for active shooter drills. (A depressingly American sentence.) 'Picher was a town of 20,000 people. It had three hospitals and a theatre and a JCPenney's,' Summer says, incredulous. 'At one stage it was the biggest town in north-eastern Oklahoma.'

We turn down what would have been a residential block. Here and there, daffodils mark out rectangles in the undergrowth: the ghosts of old gardens. Fire hydrants stand on vanished street corners. Floodlights loom over an overgrown ballpark where kids once played in the shadow of the chat piles. 'What do you think that the siblings and the other kids are doing between games?' Summer says, shaking her head. 'Playing in the chat.'

The next day, I go back to Picher alone, and wander through collapsing bungalows and houses being ripped apart by rising firs. Prairie grass is reclaiming the pavements and yards. Songbirds burble loudly on the power lines. After a while, I notice a vulture is trailing me high overhead, perhaps curious to see a human being. The last remaining houses are shells, their windows and doors having been salvaged along with everything else of value. On some, KEEP OUT is scrawled on the walls. I soon see why: the ceilings have collapsed, revealing nests of insulation. Asbestos, probably. Inside the kitchen of one house, I find a manual for a Sunbeam oven, though the oven itself has been taken. It's hard not to imagine someone chopping vegetables on this exact Formica counter, not that long ago, as children rode their bikes in the street. These houses (for in the absence of people, they are no longer homes) are now technically waste, too. Who will dispose of them? If not people, then eventually the wind.

Before I leave Picher, Summer drives us down to Tar Creek itself. It's quiet. We park on a road bridge downstream from two chat piles. We climb out of her truck to find someone has dumped a deer carcass on the roadside, surrounded by bags of trash: dog food, Marlboro red cartons, a lone leather boot. 'Someone went hunting,' Summer says, irritated. She finds worse. The sheer size of this place, and the regularity of sink-holes, means that her team inevitably find bodies.

The bridge marks the confluence at which Tar Creek joins with Lytle Creek, a stream which is no longer really a stream. 'At this point it's one hundred per cent mine discharge, so it's normally orange,' Summer says. But the storm earlier in the week has swelled the river, so today the water is more of a muddy reddish-brown. Summer points into the trees. A few yards away, a pipe is spewing water into the air like a fountain. 'All of this stuff is coming up out of a mine collapse, and becoming surface water. So the first thing that happens is the iron falls out, oxidises, and that's how it turns orange,' Summer says. She gestures downriver. 'It'll stain everything it hits from here to Grand Lake.' When surface water enters the mine, it reacts with the exposed ore to form sulphuric acid.[34] This acidifies the stream water, killing marine life. The optimum pH for water like this is around 7.4. In the 1990s, a surveyor for the Oklahoma Water Resources Board measured water from Tar Creek with a pH as low as 1.7. 'You still see a few fish,' Summer says, 'but if they get this far upstream, they don't live.'

Acid mine drainage has been called the 'global environmental crisis you've never heard of'.[35] In the US, which is estimated to contain at least 500,000 abandoned mines,[36] 189 million litres of mine water escapes from former mines every day.[37] The exact contaminants contained within it depend on the material being mined, but can range from heavy metals to gold, mercury, arsenic, and radioactive wastes. The problem is chronic in former mining countries; in South Africa, for example, waste from former gold and uranium mines has polluted drinking water in a country already mired by drought. Once begun, acid mine drainage is notoriously difficult to stop. In 2015, an EPA cleanup crew at the Gold King Mine in Colorado accidentally released 13 million litres of mine water containing arsenic and heavy metals into the Animas river, staining it mustard-yellow.[38]

Flooding is a constant risk. 'We have entire chat piles here in the floodplain,' Summer says, as we head back to the truck. 'I've seen a five-hundred-year flood, and I've only been here six years.' The chat piles are so massive and absorb so much water, she says, that removing them is actually increasing the flood risk downstream in towns like Miami. 'So it's also a climate change problem.'

On our way back to Quapaw, Summer has one more thing to show me. As we're driving back past the chat piles – excavators loading hauling trucks high with stony waste – we eventually pass back out into the countryside. There, we pass what looks like an ordinary field, cows grazing in the grass. 'This used to be chat,' Summer says, keeping her eye on the road. 'We removed over a million tons from this property. It looked like a disaster.' When the chat piles are depleted, the Quapaw dig out the contaminated soil down to the underlying clay, and replace it with clean topsoil. 'Once that's done and it's clean, then we turn it back over to the landowners, they'll be able to graze cattle on it, which is what the landowners wanted.' It's beautiful, I say. What I mean to say is: it's ordinary, in the most extraordinary way. 'It will all end up looking like this,' Summer says, though that's not quite true. The chat within the repositories will remain as hills, visible above the plains for miles, a permanent topographical scar in the landscape. ('Many of the tribes here were mound builders. That was our culture,' Rebecca told me, back at her office. 'We made mountains that have been there for 2,000 years. That's what I hope the Quapaw will do.')

This is how our waste outlives us: as the places we remake. Hills of garbage, hills of chat, tailings ponds. Waste as monument, waste as geoengineering. Rock, shifted unnaturally and drained of its former composition (that which we deem valuable, the Earth cares for little) will outlive even the hundred- and thousand-year timescale of landfills, so that in some distant future, if some cosmic force or tectonic break were to slice open this Earth like a birthday cake, the layers would be irrevocably altered by our presence. In Genesis, God laid waste to the Earth, and now so do we.

It is not just mining, either. The impact of our industrial waste, in truth, stretches further than the garbage lining our landfills and tailings ponds. Not just in space, but in time. Tar Creek is the waste legacy of nearly a century ago. Our actions today will carve a similarly deep impression. Take fossil fuels. As the Earth moves slowly – and hopefully, ever more quickly – towards low carbon energy, our vast fossil fuel infrastructure is decaying. Oil rigs. Pipelines. Power plants. Great seafaring tankers, destined to be carved up on some poor and unsafe beach in Pakistan or Bangladesh. Billions of cars. Every single piece of our fossil-fuelled world will eventually need to be dismantled, the soil left behind remediated, the waste that cannot be recycled interred somewhere safely in the earth. We are not even approaching readiness for the scale of the challenge. Will we rise to it? Or will we, too, leave it to our descendants to clean up our mess?

I'm still thinking about legacy on the drive back to Quapaw. Some of the workers on the chat piles at Tar Creek are now third and even fourth generation, Summer says: sons and daughters cleaning up the mistakes of their fathers and grandfathers. Summer, at least, is optimistic. In 2021, President Biden signed a $3 trillion infrastructure law that reinvigorated Superfund after years of budget cuts and decline, and reinstated part of the 'Superfund tax' for polluters. As a result, the EPA has reallocated $16 million per year to finally finish the job at Tar Creek. Even then, the rest of the cleanup is at least a decade away. 'Tar Creek wasn't created in a day, and it won't be cleaned up in a day,' Summer says, as the chat piles recede in the rear-view mirror. 'And it will probably take us every bit of sixty-five years to clean it up.'

12

HAZARD

'Don't say Jenga'

An alarm sounds, rhythmic as a heartbeat, clarion clear. My own pulse quickens. An alarm is the last thing you hope to hear when standing inside a nuclear facility.

'It's the plant monitoring system,' Jonathan Clingan, a plant manager at Sellafield's vitrified nuclear waste store, yells over the din. 'It's telling you that it's operating – so if it goes *off*, there's a problem.' I take a brief moment to imagine what might cause a nuclear facility's radiation monitoring system to fail, before we hurry on. Sellafield isn't a place I want to linger.

It's winter, and I've come to see an ageing nuclear facility in Cumbria, on England's north-west coast. It's here, on a thin sliver of coastline between the Lake District national park and the Irish Sea, that Britain processes and stores the nuclear waste is has generated since the dawn of the nuclear age – to date, more than 80,000 tonnes of it.[1]

Nuclear waste is unlike other wastes. It is not only the danger – nuclear waste emits alpha- and beta-particle radiation, as well as gamma rays, with the potential to corrupt cells and smash apart DNA, causing mutation, cancer, haemorrhage, and eventually death – but the timescale. Trash inside a landfill might decay over decades, plastic over hundreds or even a thousand years – the truth is we don't know yet. But the half-life of Plutonium-239 created inside the reactor cores of nuclear power plants is 24,100 years. Uranium-235, the fuel used to power the reactors,[2] has a half-life of 700 million years. To dispose of nuclear waste is to think in geological time.

Uranium is older than the Earth, forged more than 6 billion years ago by exploding supernovae and colliding neutron stars.[3] It is, by any measure, a miraculous element: a single pellet barely larger than a multivitamin can generate as much energy as a ton of coal, without any direct carbon emissions.[4] But, as with everything we plunder from the Earth, getting at it creates waste from the moment of extraction. To manufacture nuclear fuel, uranium must first be mined as an ore, then crushed, dissolved – often in sulphuric acid, itself toxic – and concentrated into urania, better known as yellowcake. The yellowcake (which, it turns out, is actually kind of brownish) is then further enriched to produce pellets of Uranium-235, thousands of which are stacked inside the fuel rods that power a nuclear reactor.

The process is dirty, dangerous, and inefficient. A tonne of unprocessed ore can yield as little as 0.2kg of uranium.[5] The ore itself emits radon gas, which causes lung cancer and leukaemia, and the mining process generates enormous quantities of sludge and liquid laced with radioactive material, which must be stored, usually in vast tailings ponds. The greatest radioactive disaster in American history is not, as most assume, the 1979 reactor meltdown at Three Mile Island, but the breach of a uranium tailings dam in Church Rock, New Mexico, four months later. The catastrophe released 360 million litres of radioactive water and more than 1,000 tonnes of uranium sludge onto the

Navajo Nation, poisoning its drinking water, and leaving a toxic legacy that is still being felt today.

When fuel rods are exhausted, they must be removed from the reactor core and replaced. The spent fuel, still emitting harmful radiation, must be disposed of. So too must anything that has come into contact with it, because if you expose anything to strong radiation for long enough, it too becomes radioactive. This can range from protective clothing to the fuel casings, and eventually the core itself, even the very building that surrounds it. The spent fuel can be reprocessed, and around 96 per cent of it recovered into reusable uranium and plutonium isotopes; unfortunately the remaining 4 per cent consists of some of the most dangerous material ever created by mankind. All of this waste must be squirrelled away from human life, from *all* life, and hidden somewhere where it can decay until safe again.

In the 1940s, as the warring superpowers raced to harness the astonishing potential of nuclear fission, little thought was given to the waste that would be left behind. At first, the nuclear industry's solution was simple: dump it in the ocean. Between 1946 and 1993, when the practice was finally banned, the world's nuclear powers tipped hundreds of thousands of tonnes of radioactive waste into the sea, packed tightly into steel drums or discharged as wastewater. That which wasn't flushed away was simply buried, or covered over where it stood. The Marshall Islands, for example, are home to the Runit Dome, a colossal concrete sarcophagus built by the US military to house the 73,000m³ of irradiated soil created by weapons tests during the Cold War. Known locally as the Tomb, the dome is built atop a 100m-wide blast crater on the Enewetak Atoll, and holds not only an unknown quantity of plutonium, but also radioactive soils illegally shipped to the Marshalls from Nevada and dumped there in secret, without approval from the island's government. The Dome has been leaking for years, and is at risk of being swallowed by the rising sea. However, despite an international tribunal in 1988 ordering the US government to pay $2.3 billion to help with the cleanup, it has so far refused.[6]

In the years after the war, scientists concocted various ideas of how to deal with high-level nuclear waste – some, such as packing it into rockets and firing it into deep space, less workable than others. Almost all experts now agree that the only practicable long-term storage method is geological disposal: burying the waste deep underground inside a carefully engineered facility where it can be isolated from the elements for millennia to come. However, despite several countries having discussed geological disposal since the 1970s, the only one currently even close to completing such a facility is Finland. And so, with nowhere to go, most of the world's nuclear waste is still sitting inside a handful of surface-level 'interim' facilities, including Mayak, in Russia; La Hague, in France; and Sellafield, waiting for its creators to decide its fate.

I arrive early. Until the 1940s, Sellafield was farmland, and the surrounding countryside is still chequered by hedgerows and flecked with sheep grazing idly in the rising light. What was once the old stone farmhouse now makes for a rather incongruous welcome centre, with gleaming cutaways of nuclear waste containers displayed proudly in the lobby. At its peak, 1,000 visitors a day travelled from around the country to visit Sellafield's nuclear plants and to learn about the dazzling potential of atomic energy. But after 9/11 the tours stopped due to security fears, and never resumed. Today the site is kept under the highest supervision, ringed with barbed wire fences and staffed by its own armed police force, the Civil Nuclear

Constabulary, who stand vigilant at the entrances and fence lines. Just getting here has taken me eight patient months of interviews and security vetting.

I'm to be accompanied by Ben Chilton and Corrin Nichols, two of Sellafield's communications team, who have been assigned to make sure I don't get into trouble. Ben (early forties, bearded, polite) and Corrin (dark hair, kind, with a Cumbrian accent so thick and brisk that I find it hard to keep up) have both worked here for more than a decade, and know every in and out of the site. After watching a brief induction video – which reassuringly treats the risk of a nuclear accident with the cheerful detachment of an in-flight safety briefing – we climb into a minivan for a tour.

Like so much of nuclear history, Sellafield began with the bombs. During the Second World War, Britain cooperated with the United States on the Manhattan Project, which designed and built the nuclear weapons used at Hiroshima and Nagasaki. But in 1946, with the Cold War brewing, the Americans withdrew from the collaboration, and the British government decided that it needed its own nuclear weapons programme. Sellafield, then known as Windscale, was chosen as the site for a new plant intended to refine uranium into weapons-grade plutonium. The location was perfect: a remote former munitions factory, it was close to the sea, and downriver from Wastwater, England's deepest lake. By 1951 Windscale's two reactors, Windscale Piles 1 and 2, were producing the Plutonium-239 used in Britain's first nuclear weapons tests. Soon Windscale was the burgeoning hub of Britain's nuclear industry.

To wander around Sellafield is to be transported back to a bygone era, where nuclear power represented a utopian future. In 1956 the world's first commercial nuclear power plant,

Calder Hall, opened here, providing electricity to the national grid until its eventual closure in 2003. That was followed in 1962 by the Windscale Advance Gas Reactor, the prototype on which eight of Britain's nine remaining nuclear reactors are based, housed inside a spherical white building affectionately known as 'the golf ball'. That was then. The last power plants on the site have long since shut down; only their husks remain. (Ironically, Sellafield now draws its energy from a gas power plant built just outside its limits.) Calder Hall's cooling towers – the looming concave structures you'd normally associate with nuclear power stations – were demolished in 2007. Windscale Pile 2's chimney came down in 2001, and work is now proceeding to demolish the ruins of Windscale Pile 1. The effect is that the once bold skyline of Sellafield is slowly diminishing, the entire site slowly withdrawing into itself.

Today Sellafield is less a workplace, and more like a small town. More than 10,500 people work on the site, which has its own hospital, post office, and railway station. 'Sellafield pretty much *is* the local economy,' Ben says. The plant is so old, and employment here so stable – nuclear disposal being a boutique skillset – that many of the workers are lifers, and in some cases multiple generations of the same family work together, even in the same building. More than once as we're driving around the site, Corrin waves at a passerby, saying, 'That's my cousin!'

We arrive at the former Thermal Oxide Reprocessing Plant (THORP), a nickel-grey building big enough to swallow a ferry, and ride up a small elevator to the viewing gallery. This would once have been a stop on the tour, but now is strictly invitation-only. Even then, thanks to the pandemic, barely anyone has been here in months. The lights are off when we arrive. Corrin hits a switch, and the room flickers into view. It's a long, joyless space, with floor-to-ceiling windows looking down on the reprocessing line, while along the back wall a pristine photography exhibit celebrates the history of Sellafield.

Sellafield is unusual for a nuclear facility, in that it is also the site of a nuclear disaster. In 1957, a fire broke out inside Windscale Pile 1. The reactor burned for three days. Although nobody died – an innovative filter atop the Pile 1 chimney prevented much of the worst radioactive material from escaping – iodine, caesium and polonium were released into the skies over Cumbria, and drifted across the countryside as far as Europe.[7] Radioactive dust tainted nearby farms. Animals were culled; milk from local dairy farms was flushed into the sea. Although the British government concealed the true extent of the fire for decades, studies have since connected the radioactivity released to as many as 240 cancer cases, making Windscale among the most deadly nuclear incidents in history. (One reason that the site was renamed Sellafield in 1981.) The reactor core itself, too radioactive to save, was more or less abandoned after the fire. The remains of Pile 1's chimney were sealed and left for decades. Although work to decommission the facility began in the 1980s, it is expected to continue until at least 2037.

Today, Sellafield's primary purpose is decommissioning – atoning for the mistakes of our nuclear past. During the height of the Cold War, when Piles 1 and 2 were operating, the waste created was disposed of using crude and reckless methods. Once removed from the reactors, nuclear fuel and intermediate-level nuclear wastes are still highly radioactive, so must be placed into large pools of demineralised water, called storage ponds, which soak up the excess radiation and heat. The early storage ponds at Sellafield were crudely designed, little more than open-air pools; gulls, unaware of their contents, would fly down and bob happily on the water, soaking up radioactivity. (In 1998, more than 1,500 birds living around the site were found to have absorbed so much radiation that they were culled and their corpses disposed of as low-level nuclear waste.)[8] But the true hazard lies under the surface. During the 1972 miners' strike, the UK's fleet of Magnox nuclear power stations worked overtime to keep the national grid running.[9] The excess waste generated was too much for reprocessing facilities to handle, and so Sellafield's plant operators at the time simply dumped it in the storage ponds, where it settled in layers of radioactive sludge. 'The water, which was supposed to be a radiation barrier, became a radiation hazard in its own right,' Corrin explains.

The spent fuel was tipped inside two huge concrete silos at Sellafield and essentially left to decay. So too was the metal cladding from Magnox fuel rods. The Magnox silo and storage pond at Sellafield, still stuffed with highly radioactive material, are now considered the two most hazardous buildings in Europe (and therefore, to my dismay, not included on my tour).[10] Over the years the slow assault of the elements and corrosive sea air have left the buildings cracked and deteriorated. Both are now in the midst of a prolonged and expensive cleanup programme.

As the waste is still emitting radiation, the removal is done remotely using robots, which transfer the waste into new containers to be stored inside a nearby vault. 'That one there is a real success story,' Ben says, pointing out the Pile Fuel Storage pond. 'We're getting to the stage now where we're possibly going to be sending in specialist nuclear divers to finish it off.' Magnox, on the other hand, is still off-limits. 'You won't be sending divers in there any time soon.'

There are more than 200 radioactive buildings at Sellafield. Some are remaining sections of the decommissioned plants, but most are now used for various levels of waste storage. Radioactive waste is generally classified into different categories: material that has been only lightly irradiated and so emits fewer than 2 millisieverts per hour* is dubbed 'low-level waste'.[11] This is typically placed inside drums, compressed, and buried in sealed landfills. In the UK, this happens a few miles south of Sellafield at a government-designated Low Level Waste Depository in Drigg. At its peak, Sellafield sent around fifty containers of low-level waste to Drigg by rail each week. Now, thanks to aggressive waste-reduction targets, only thirty or forty are sent every year. Today, however, the compressor used to crush the low-level waste containers down (to save space) is broken. 'It's an old plant, you know,' Mark Rothery, a friendly, curly-haired engineer who runs the low-level waste compaction facility, tells me. 'We do suffer from mechanical faults, degraded parts, obsolete parts – it's not different from any other ageing facilities,' he says. 'That's what we get paid for, to fix these problems.'

Intermediate-level wastes, such as fuel rod casings and the machinery from inside the reactor core, emit high levels of radiation but little heat. These are chopped up and poured into 500-litre stainless steel drums, where it is mixed with cement, placed inside $3m^2$ boxes, and stacked up inside highly shielded warehouses.

'Think about an IKEA without anything in it,' Corrin says.

Ben gets out his phone, and pulls up a picture of a vast warehouse. 'The picture doesn't do it justice,' Ben says. 'It is absolutely *massive*. We went last year—'

---

* There are a lot of ways to measure radiation. Sieverts refer to the received dose in human tissue. The UK estimates the average person is exposed to 2.7 millisieverts in 'natural' radiation per year.

'Got a selfie.'

'—And the acoustics! Make a noise, and it goes on for about ten minutes.'

Sellafield's intermediate stores are designed to hold waste for no more than one hundred years, by which time the UK government hopes to have constructed a long-term Geological Disposal Facility (GDF). But any hope of that actually happening is still so remote that the waste stores were built with no export facility – that is, you can put waste in, but there's currently no reasonable way to get it out. For a supposedly temporary solution, it feels awfully permanent. 'Until we have a GDF, this is what we're doing with it,' Corrin says.

We move along the viewing gallery, looking down on the reprocessing line, a series of towering bays and deep pools veined with pipes of unknown origin and terminus. A yellow beam crane hovers overhead, ready to snatch up waste canisters like an arcade slot machine. Across the void I can see one man in a control room, his face lit by his computer monitor, but otherwise there are no humans in sight, giving the space an eerie stillness.

The THORP reprocessing plant opened in 1997. It was originally intended to recycle fuel from both the UK and other countries that lack their own reprocessing capacity – Japan, Germany, Australia, Switzerland, the Netherlands. But the plant barely reached a fraction of its intended capacity, and was beset by delays and technical problems. In 2005, workers at THORP discovered a ruptured pipe had been spilling tens of thousands of litres of waste solution – containing an estimated 20 tonnes of uranium – into an emergency containment tank, a fact that had gone unnoticed for nine months.[12] Even when THORP was working, reprocessing is so expensive, and virgin uranium so cheap, that it struggled to compete with newly mined fuel from Kazakhstan, Australia, and Canada. 'We produced good quality fuel, but there wasn't demand for it,' Corrin says, sadly.

But the real death knell for reprocessing was the Fukushima nuclear disaster in 2011, after which Japan, Germany, Switzerland and Belgium announced the mass closure of their nuclear plants. Almost overnight, the market for reprocessed fuel collapsed. In 2018, THORP halted reprocessing, decades ahead of schedule; Sellafield's Magnox reprocessing plant has also closed. Instead, THORP's cooling pools are now being used to store waste – at least until 2075, at which point the plant itself is set to be decommissioned and dismantled.

At THORP, the spent fuel arrives in transport flasks – towering white containers, that look a bit like gas canisters. It must first be repackaged into storage flasks, which are then loaded four-at-a-time inside large steel crates and lowered into the storage ponds. The ponds are a delicate environment; at this point the waste is still hot, and so fresh water must continuously be flushed through the system, to stop it from boiling off. The water must also be kept at a stable temperature and pH, and any mineral content removed, to prevent the containers from corroding.

The end of the gallery overlooks a gigantic pond: 144m long, and 8m deep. The water is the blue of unwashed jeans. The waste containers, stacked three high, are visible under the surface, like some strange man-made reef. 'It's just a game of Jenga really,' Corrin, says, cheerfully. I picture a tumbling stack of nuclear coffins.

'No, not Jenga!' Ben says.

'Not Jenga.'

'Don't say Jenga!'

The collapse of the nuclear industry in Europe and Japan

(China is still building nuclear plants at pace) has also created new questions about the waste at Sellafield – for instance, what will ultimately happen to all of the reprocessed fuel. In addition to the medium- and high-level waste stored here, Sellafield is currently sitting on around 90,000 tonnes of reprocessed uranium and 140 tonnes of civil plutonium, the largest stockpile in the world.[13] For context: as little as 4kg of plutonium is needed to build a nuclear weapon. An academic once estimated that the store of plutonium at Sellafield – which is so heavily guarded that my guides won't even tell me where it is – would be enough to construct 20,000 atomic bombs.[14] But without any buyers, it's just sitting there.

'It's semantics,' Ben says. 'It's how you categorise that fuel flow. Is it an asset? Is it a waste? In the States, largely that's the end of the process. They call that used fuel "nuclear waste" and that gets buried. So do you class that as a waste – which it *is*, it's a by-product that probably has no practical use. But it's still classed as an asset by the government, for use either in a reactor or . . .' He tails off, leaving the alternative unspoken.

We leave THORP and head a little further up the road, until we arrive at the place that I've come to Sellafield to see: the Vitrified Waste Store. Jonathan Clingan, the plant manager (whom we met at the start of this chapter), meets us outside. Technically, Jonathan's job title is Highly Active Liquor Programme Manager, which sounds a lot more fun than it is. An unassuming, grey-haired engineer in a North Face fleece and black glasses, Jonathan has worked at Sellafield since 1986, meaning he has worked here since before I was born. 'I've been here so long I see the decay,' he says – that is, he has seen the radiation fall. Jonathan started out as a mechanical engineer, repairing broken things, and over the years has worked almost everywhere at Sellafield, seen every part of the process. He's unfailingly polite, but his facial expression and walking speed suggest that he has better things to do.

When spent fuel is removed from a nuclear reactor, it is a hot mess. The chain reaction leaves behind a hodgepodge mass of decaying elements, which are still firing off neutrons and radioactive particles like sparks from a roman candle. To recover the depleted uranium and plutonium inside, the fuel must be dissolved in nitric acid, and put through a chemical separation process. The radioactive by-products created in the heart of the reactor are left behind, in the form of 'highly active liquor'. This material – including americium, strontium, and other rare and highly emitting isotopes – is classified as 'high level' nuclear waste. A single canister can emit 2,000 sieverts of radiation per hour. (Around 5 sieverts is enough to kill you.)

To stabilise it, the liquor is mixed with sugar and molten glass, *vitrified*, before being poured into dedicated storage canisters, and then interred in a shielded vault, far from prying eyes. This is Jonathan's job: to protect humanity from the worst of what it has made.

The vitrification line is a high-hazard area, which requires additional safety measures. We pass through a changing room, where I'm given clean, bright orange socks and a blue lab coat to wear. The coat itself has seen better days: the patch on the breast pocket still reads BNFL, for British Nuclear Fuels Ltd, the state-owned company that used to run Sellafield but has been defunct for more than a decade. Jonathan hands me a safety helmet and a pair of steel-toed safety boots with the toecaps painted bright red. This is ostensibly to remind plant workers to remove their footwear before leaving the facility, lest they track radioactive material out of the site. In reality, it creates the awkward feeling of wearing clown shoes, which along with the outdated outerwear and ill-fitting helmet, does not leave me brimming with confidence.

We pass through another set of guarded, security-coded gates. Then, under the watchful gaze of two guards (due to the pandemic, even security is socially distanced) and I am handed one of those wand-like devices that you see at airport security, used to detect explosive chemicals. I sweep it over my clothing like a lint brush, sheepishly, before the guards wave me through. Once we pass through I hear the sound of the plant's radiation monitoring system, its rhythmic *blip-blip-blip* evocative of a hospital ward.

Jonathan hands me a dosimeter, a small white device used to measure radiation. 'That will monitor your dose in real time. If a dose is too high, or if the accumulated dose is too high it will go into alarm,' he says. I glance at the screen, which thankfully reads o, and slip it into my pocket. 'My expectation is where you're going you might hear a few beeps,' Jonathan says. 'But don't worry.'

I ask Jonathan what people say when he tells them that he handles nuclear waste for a living. 'We tend to not divulge where we work,' he says. 'In Cumbria you might.' The locals understand. But outside? It only provokes questions. Jokes. Uneducated assumptions. Plus, there's the security considera- tion. 'It's not like I'm hiding. If somebody said, "What do you do?" I'd say a programme manager or a mechanical engineer.' If they ask where he works? 'I say, "At a big factory."'

We pass down an empty corridor. Through a window, I can see into a control room where a handful of plant workers, all men, are monitoring the vitrification line. One of the strangest things about Sellafield is how anachronistic it feels – all boxy control panels with square illuminated buttons cast in retro shades of eau de nil. It's like being on the set of a Cold War-era spy movie. Even the modern computers all appear to be running an operating system long out of date. This, of course, is intentional – the systems might look old, but they're built for security, longevity. Still, I can't shake an unexpected feeling that all this is more precarious than it seems.

We walk on. A series of windows look onto the vitrification line, which is cast in a yellow light. At various points along the corridors manipulation arms, used to control the waste-handling robots, jut down from the ceiling. The windows are leaded glass, several feet thick, but lensed so that everything looks close enough to touch. Inside the line are pieces of machinery and other broken things, now radioactive waste, which cannot be removed until the next plant shutdown. I put my fingers against the pane, my brain unable to reconcile the certainty that I am perfectly safe, and yet untold danger is just a few feet away. It's a familiar feeling, although the reason doesn't hit me until a few weeks later, on a trip to the zoo with my daughters. It's the sensation of confronting a tiger in a cage.

For the liquor to be stabilised, it is first passed through a series of evaporators. This concentrates the liquid, while drawing off some of the most volatile elements. 'You're left with what looks like coffee granules,' Jonathan says. 'I know that because I've seen it inactive. I haven't seen it *active*, otherwise I wouldn't be here.' These granules are then mixed with molten glass and sugar. 'Tate & Lyle sugar,' Jonathan says. (Other brands apparently work equally well.) The sugar sticks the volatile molecules together. The plant recently switched to a new glass mixture, which is both more compact as well as more stable.[15] 'We'll save hundreds of containers,' Jonathan explains. Once the material is vitrified, it is poured into a storage flask, which is later nested into another container like Russian dolls.

At the end of the line, the containers pass into their temporary resting place: the Vitrified Product Store, an immense hall with several hundred storage shafts sunk into the floor, more than 10m deep. 'This is where the containers will be stored for the best part of this century,' Jonathan says, his voice echoing. The shafts are sealed with steel plugs more than a metre thick, and topped off with yellow caps about the size of a manhole cover. A few have signs taped over: 'floor plug damaged – do not use'. I glance at my dosimeter, but it's still at zero.

'You can feel the heat,' Jonathan says. I press my hand to the floor. Although there are several metres of metal and concrete shielding me from the waste, the plug is indeed slightly warm, like a mug of tepid coffee. This chamber has no mechanical ventilation system (an electrical failure would be too much of a risk) and so the cooling relies on old-fashioned convection currents. The more radioactive the material the greater the heat, but the shorter the half-life, and so the waste must be moved around as it decays and cools. 'You want to start close to the inlet initially, because that's going to start drawing the air through. Then you want to start spreading the heat, so we don't have a cold spot in the vault,' he explains, as if we're talking about cooking on a barbecue. Monitoring stock levels, moving things around, keeping everything cool – you're like the country's most overqualified stockroom manager, I joke. Jonathan doesn't laugh.

Currently there are just over 6,000 containers already in the vault, with room for another 2,000. Some of them are being held for international clients – the Japanese, the Germans, the Swiss, Italians, and the Dutch – who pay Sellafield to temporarily store the waste until it can be sent back to its country of origin on one of the UK's nuclear cargo ships. To date, Sellafield has returned around 900 containers. 'We actually just had some intermediate-level waste that we shipped back to the Australians last week,' Jonathan says. My mind briefly conjures images of customs paperwork and ocean storms.

The biggest share of foreign waste is Japanese. However, following the Fukushima disaster, shipments back to Japan are currently on hold. What would happen if the Japanese were to simply refuse to take it back? Jonathan thinks for a moment. 'It would be the cascade effect,' he says. 'So if the Japanese said they wouldn't take it back, then the Germans don't want it back, the Italians say they don't want it back – then we have a capacity issue.'

I wander around the vault, the lemon-yellow covers polka-dotting the floor, trying to imagine the vast energy stored beneath me. Such is the concentration of radioactivity at Sellafield that it was once estimated that a major fire or explosion, such as one caused by a terrorist attack, could generate a fallout between ten and one hundred times greater than at Chernobyl.[16] The resulting plume of radiation would render the surrounding countryside unliveable for decades, and require evacuations as far south as Manchester. Accidents at nuclear waste storage facilities are not without precedent: in 1957, the year of the Windscale fire, high-level nuclear waste at Russia's Mayak nuclear facility near Kyshtym overheated and exploded, causing a massive airborne release of radioactive material, which drifted north over the Urals for thousands of miles. Eleven thousand people were evacuated, and their villages abandoned. But the Soviet government ordered the disaster hidden – even those being evacuated were not told why – and so the true impact of the Mayak disaster is unknown, with long-term fatality estimates ranging from the tens to the tens of thousands.[17]

In 2014, a fire broke out inside the Waste Isolation Pilot Plant (WIPP), an underground facility in New Mexico where the

United States stores the high-level waste from its nuclear weapons programme.[18] Plutonium and americium were blown into the air above ground, and twenty-two workers were exposed to radiation (none, thankfully, to serious levels). The cleanup cost $500 million, and the facility was forced to close for three years.[19] It was only later determined that the fire was caused by contractors packing kitty litter into the containers to soak up moisture. Rather than the correct kind of kitty litter,* made of bentonite clay, one subcontractor had used organic kitty litter, triggering a chemical reaction inside the containers. Such an event is vanishingly unlikely at Sellafield, with its modern technology and stringent safeguards. But such accidents serve as a reminder of the waste's inherent danger, and the urgency of finding it a permanent resting place.

Sellafield's vitrified waste plant opened in 1990. It was initially designed with a lifespan of fifty years; it was widely assumed that the UK would have built a Geological Disposal Facility by then. As it is, that deadline will not be met. In the UK at least, a GDF is still decades away. 'We're looking at towards the turn of this century,' Jonathan says. His team regularly inspects the vaults for cracks and signs of ageing; the hope is to eke out this facility for as long as possible, to avoid the expense of building another interim store.

The truth is that even after decades of wrangling, the UK has never even come close to building a geological disposal site. The political opposition is simply too great. The last serious attempt stalled in 2013, after a plan to pursue two feasibility studies in Cumbria was fiercely protested by environmental campaigners,

---

* Yes, apparently there is a correct kind of kitty litter when you're dealing with nuclear waste.

who argue that the local geology was unsuitable and the risk of pollution too severe. Incredibly, the showdown over where (or in this case, where not) to store the UK's nuclear waste for the next hundred millennia came down to a county council meeting in Carlisle that January, where after a tense discussion, the councillors voted against, seven to three.[20]

The UK is not the only country that has struggled to build a permanent disposal facility. The United States had been pursuing a site at Yucca Mountain, in Nevada, since 1987. However, despite spending more than $9 billion on the project and digging an 8km-deep exploratory tunnel into the volcanic tuff, the project was halted by the Obama administration in 2011, and has since been abandoned.

Only one country, Finland, is close to opening a geological disposal facility. Known as Onkalo (Finnish for *cavity*), the site is situated on the country's south-west coastline, deep in the birch and pine forests around Olkiluoto, Finland's largest nuclear plant. Onkalo is the designated resting place for the high-level waste generated by Finland's nuclear plants – eventually, around 6,500 tonnes of it. The waste will be stored inside copper canisters, sealed inside grout and cement, and then placed inside an underground tunnel network up to 70km long that is being dug deep down into the granite bedrock. There it will stay, in concrete tunnels up to 520m deep, for at least the next 100,000 years.

We are not built to think in such timescales. Human beings – lucky ones – might survive a few decades, empires a few centuries, religions a few millennia. The oldest known cave art dates back 46,000 years or so, the oldest writing just 5,600. A hundred thousand years ago the Earth was still in the grips of the last Ice Age, homo sapiens scraping for survival alongside Neanderthals and giant ground-sloths. But nuclear waste forces us to think in deep time; not in lifetimes, but in half-lives. The waste buried at Onkalo will remain radioactive for tens of millennia. Which forces questions, like: how do we build something to last longer than anything in human history? And, what do we tell our descendants about the danger we have left for them?

In 1980, the United States Department of Energy gathered a group of linguists, archaeologists, psychologists, and designers to consider this problem. Dubbed the Human Interference Task Force, the group was tasked with designing a warning message that might endure for just 10,000 years – the operational life of the (now abandoned) Yucca Mountain Depository. To run the Task Force, the department appointed Thomas Sebeok, an esteemed Hungarian semiotician, and a call for proposals was put out in the *Journal of Semiotics*. An eclectic mix of individuals sent in their ideas, some more fantastical than others. Many coalesced around images such as skulls, which it was thought would last beyond any written language. But it turns out that even ideas we might take for granted, like *skull = danger*, do not always translate across cultural boundaries – and you only have to look at Egyptian hieroglyphs to know that pictures are not always easily translated when divorced from their cultural context.

Another idea was the telling of folk tales, like those of the great Icelandic sagas, which have been passed down in fairly consistent form over centuries. A Polish science fiction author proposed launching satellites that would broadcast the locations and risks of the nuclear waste from orbit. (One problem: you'd need to replace the satellites periodically, not to mention it assumes the continued existence of satellite communications.) Sebeok himself proposed the creation of an 'atomic priesthood', which would pass down the history and science of nuclear waste through 'artificially created and nurtured ritual-and-legend'. In doing so, Sebeok argued, the priesthood would create enough 'accumulated superstition to shun a certain area permanently'.[21]

My personal favourite is the proposal put forward by Françoise Bastide and Paolo Fabbri, two semioticians based at EHESS* in Paris. They proposed breeding cats – language being fleeting, but our love of pets timeless – whose fur would change colour or glow when exposed to radioactivity.[22] The essential role of these 'Ray Cats', Bastide and Fabbri suggested, could be preserved and passed down in proverbs and myth, creating a kind of glowing cat religion. (You get the impression they weren't dog people.) Curiously, the idea has had a minor resurgence in recent years, thanks to the internet and the arrival of gene-editing technology – to the extent that a Canadian biotechnology startup, Bricobio, has declared an interest in attempting to breed one.

Evidently not quite convinced by the cat plan, in 1991 Sandia National Laboratories convened another group of experts to design a warning system for the Waste Isolation Pilot Plant in New Mexico. The Sandia group split into two teams: one tasked with engineering physical barriers, and another concerned with what messages they might convey. One idea was to erect a 'landscape of thorns' above the site, to physically deter intruders. (A minor flaw being that humans love a challenge.) Several designers suggested using images of wastelands or wounded bodies. Ultimately, the panel's solution was to surround the site with thirty-two huge granite monoliths. Each will be inlaid with warning messages in English, Spanish, Chinese, Hebrew, Navajo, and Latin, and intentionally horrifying images based on

* École des hautes études en sciences sociales.

Munch's painting *The Scream*. These monoliths will surround a massive earth berm erected over the burial site. Inside the berm will sit an empty room, its walls covered in a series of warnings. The message itself, the scientists suggested, should sound something like this:

This place is not a place of honor . . .

No highly esteemed deed is commemorated here . . . nothing valued is here.

What is here was dangerous and repulsive to us. This message is a warning about danger . . .

The danger is still present, in your time, as it was in ours.

The danger is to the body, and it can kill.

The form of the danger is an emanation of energy.

The danger is unleashed only if you substantially disturb this place physically. This place is best shunned and left uninhabited.

Onkalo will have no such defences. Instead, it is designed to be forgotten. Posiva, the company that now runs the site, hopes to begin depositing waste in 2023, and to continue until around 2120. At that point, the tunnel network will be backfilled with bentonite clay and concrete, and the entrance will be buried. Within a few years, any trace of it is likely to be swallowed by the encroaching pines. The gneiss bedrock into which Onkalo is carved is geologically stable, impermeable, and contains no minerals of any value – meaning it should in theory be left undisturbed for millennia, until its very existence slips from memory into myth.

If, by some strange circumstance, some future human were to dig down into the site thousands of years from now, the risk would be far lower than today. The most radioactive wastes at Onkalo and at Sellafield have the shortest half-lives; the danger will fall rapidly for the first few hundred years, before slowly stabilising into longer-lived isotopes. Should Onkalo in fact survive in 100,000 years, the risk to whoever might discover its remains will not be much greater than getting a CAT scan. After a few million, the risk will be indistinguishable from background radiation. Even so, traces of the nuclear waste left behind will still be evident in the geological record billions of years from now, as perhaps the most enduring marker of the Anthropocene.

My Sellafield tour is over, and so we drive back to the farmhouse, talking about Onkalo. Lately, there is renewed hope that the UK is getting close to securing its own geological disposal facility. In 2019, the government announced it was resuming plans to finally build one; at the time of writing, several sites are currently being considered, including two in Cumbria, not far from here. 'They've tweaked the process, so that they don't need the county council any more,' Ben says. (So much for local democracy.) Even so, any construction is still at least a decade away. First there are geological surveys to complete, environmental impact assessments, community outreach – the long and slow manufacture of consent. One hundred thousand-year decisions are not to be taken lightly.

Still, driving around the cracked and weathered buildings, it's clear that a decision must be made soon. Sellafield is getting old; so too are many of its most experienced engineers. The waste here cannot be held in purgatory for ever. 'I think that something that people missed last time around is that the waste is mostly here already,' Ben says. 'I think people thought this stuff was going to be arriving by the truckload through the Lake District National Park. It is here now.'

Nuclear power is itself undergoing a sudden renaissance. As I write this, Russia has invaded Ukraine, pushing global gas prices to record highs, and prompting fears of energy shortages throughout Europe. And with many countries racing to hit decarbonisation targets, nuclear energy is regaining support as a low-carbon and reliable energy source. France recently announced its intention to build up to fourteen new nuclear reactors by 2050.[23] Germany has delayed plans to close two of its three remaining plants, which had been winding down after Fukushima.[24] China alone has announced more than 150 new reactors.[25] And the UK has confirmed that it will build at least one, and perhaps as many as eight new reactors in the coming years.[26] It is likely that future plants will be smaller, and use alternative fuels and reactor designs that create less waste (perhaps the uranium and plutonium stored at Sellafield will be used after all). Even so, waste there will be, and so any discussion of new nuclear plants must be accompanied by plans of how to dispose of what will inevitably be left behind.

Having stood above the vitrified store at Sellafield, I'm left feeling conflicted. Nuclear power is a carbon-free energy source that could help to liberate the world from fossil fuels, which have done far more damage to this planet and cost more human lives than nuclear accidents ever have. The nuclear industry is far safer and more responsible now than it was at the dawn of the nuclear age. But it's impossible to stand so close to so much energy, so much risk – touch it, and you can feel the heat – without thinking about the burden that we are passing on to future generations. I'd come to Sellafield with questions: *What about waste that is unavoidable? Can waste ever be worth it?* But I leave feeling even less sure, the timescales too vast, the calculations too complex.

Sellafield reminded me of the container ships conveying waste to China, only in this case transporting waste not in distance, but in time. The writer and historian Rebecca Altman once coined the phrase 'time-bombing the future'. She was writing about synthetic chemicals — but really the concept was about harm, and the ways that we externalise it. What are carbon emissions, if not time-bombing the future? What is waste, left stewing in hermetically sealed landfills, or fizzing with radiation in the bowels of unmarked caves? These are a form of waste colonialism inflicted not on our neighbours, but on our children, who will bear the risks long after we have reaped their benefits. What will our descendants in future centuries think of these time-bombs — in the case of nuclear waste, literally explosive materials we deposit in the geological strata? One can only hope they will look back at our present technology the way archaeologists look at prehistory, as artefacts of a more primitive time.

My visit to Sellafield over, a taxi arrives to take me to the station. The first signs of spring are arriving, and as I board my train south a rainbow breaks out overhead, cartoonishly bright. I look out at the sun burning low over the ocean — another nuclear reactor, 146 million kilometres away, bathing the scene in radiation and golden light.

EPILOGUE

PRECIOUS

*'Consume or die. That's the mandate of the
culture. And it all ends up in the dump'*
—DON DELILLO, *Underworld*

For weeks after I get back from Sellafield, I find myself in a
rut. Throughout my journey, the scale of our waste has felt at
times both overwhelming and hopeless. Several of the people
I've met in the waste business shared similar stories: once you
really start to look at waste, you see it everywhere. It becomes
an obsession. After a time, it begins to take a spiritual toll. It
doesn't help that every time I switch on the news I see further
fuel for my climate anxiety: record heatwaves, deadly floods,
fires everywhere. The news continues to stalk me, via my
inbox and social media feeds. Chemical boundaries crossed.
Microplastics in raindrops, dust in the wind. 'Trash is the visible interface between everyday life and the deep, often abstract
horrors of ecological crisis,' Heather Rogers wrote, in 2005.[1]
What was true then feels more urgent now.

Whenever I tell people that I'm writing about waste, they
inevitably ask a version of the same question. *What should I do?*
In truth, I struggle with the answer myself. On a personal level,
my reporting has changed my life in small and large ways. It has

transformed the way I buy and dispose of things. At home, we now buy as little plastic as we can, and have swapped out plastics to degradable or recyclable alternatives wherever possible. I started using a bamboo toothbrush, and use a shampoo bar instead of shampoo from a bottle. We've ditched our clingfilm for beeswax wraps, carry refillable water bottles and coffee cups and tote bags. I spend at least as much time washing and sorting our recycling as I do the dishes. We compost.

These are not radical acts. Such substitutions have become commonplace, pushed by 'green' brands and corporations as a way to conspicuously consume our way out of climate guilt. What's worse, a lot of the time these actions are not always even better than the behaviours they replace. Cotton tote bags, for example, must be used 7,100 times before they are more environmentally friendly than using the single-use plastic bag they replace, according to a 2018 study by Denmark's Ministry of Environment and Food.[2] For organic cotton, you'd need to use the bag 20,000 times – or once a day for fifty-four years. How many of those bags will last that long? How much forest is being cut down to grow that cotton? And yet, it sometimes feels like totes are now being given out as freely as plastic. We must have at least a dozen of them at home. (The publishing industry is particularly egregious for this.) That isn't to imply that we should continue to use plastic bags, but rather that we should acknowledge that substituting one form of consumption for another without interrogating it first can lead to consequences that are just as bad, or worse.

While I've been researching this book, my social media feeds have filled up with posts and videos from what might only be described as waste influencers. (What can I say, I'm fun and the algorithm knows it.) In posts and videos with titles like 'How I Fit 5 Years Of My Trash In This Jar', these women – and in my experience they tend to be women – follow a familiar formula: they appear with a kilner jar stuffed with a tiny handful of wrappers, looking breezy and aspirational, and explain how easy it is for you to live a zero waste lifestyle, if only you follow these tips!

Zero waste is a concept with deep roots. Its current moment can be traced to the work of Robin Murray, the economist and author of the 2002 book *Zero Waste* – although Murray himself credited its ideas to Japanese industrial designers in the 1980s, who strove to eradicate waste as a means to increase quality and lower production costs. To Murray, zero waste didn't just mean not sending waste to landfill. It meant reimagining the industrial system, to be one that permitted zero toxic chemical discharges and zero greenhouse gas emissions, one that seeks to replace the linear model of take-make-waste and instead model our economy on natural cycles of breakdown and renewal. To Murray and his acolytes, zero waste – or as we now often call it, the 'circular economy' – is a radical shift, one that 'challenges the whole way we think of things and their uses, about how we define ourselves and our status through commodities, by what we cast out as much as by what we keep in.'³

The latest iteration of zero waste, which I might describe as 'zero waste as wellness', is perhaps best summarised in the writing of Bea Johnson, author of the 2013 bestseller *Zero Waste Home*, whose blogging and public persona led the *New York Times* to dub her 'the priestess of waste-free living'.⁴ I like Johnson, and find her aims noble and her methods thorough, although I found her book stifling and at times harsh. (Among her advice to eliminate waste: consider not having children, or adopting instead.) There are many other waste influencers whose ideas and energy I also find inspiring. But over time, I have become frustrated with the movement, which can often feel like a form of conspicuous consumption: overtly concerned with display, and puritanical in its attitudes. (Watching Instagram stars with their beautifully curated shelves of home-pickled goods and fastidiously Marie Kondo'd apartments, I couldn't help but think of the musky earnestness of Compost John.) That isn't to say that zero waste needn't be beautiful or aspirational. But if we are to grapple seriously with our waste, this didn't feel like the solution.

Similarly, I felt slightly unsure about the rise of refill or 'zero waste' shops, which began cropping up in large numbers around 2019.[5] These places tend to follow a familiar formula: racks of raw pulses and carbohydrate (pasta, oats, rice) ready for people to fill up their empty kilner jars, alongside carefully curated arrays of ethical (and inevitably expensive) plastic-free products, ranging from cleaning supplies to kitchen utensils.

When I started this book, there was one zero waste store near us, a twenty-minute drive away. Soon they started opening in every town, including Royston. I love zero-waste stores. They are without fail run by people who are dedicated to reducing their impact on the planet. But the specifics of the zero-waste lifestyle – the idea that we will all start carrying heavy glass jars every time we go to the store to pick up some pasta – felt unattainable to anyone who works full time or isn't wealthy. Buying goods plastic-free and in bulk is more expensive, requires time and effort, and often the use of a car to carry around all those containers. (I often noticed zero-waste shoppers arriving in Land Rovers, surely undoing the carbon benefits of their plastic-free packaging in a single trip.) As much as I wanted to embrace zero-waste shopping, the barriers to entry just felt too high. And perhaps they were: by the time the pandemic was over, most of the zero-waste shops near me had closed again.[6]

That isn't to say that refill stores won't work – in fact, they have already started to inspire wider reforms. Supermarket chains including Tesco and Waitrose have now started to introduce refill aisles for certain goods, and even plastic-polluting giants such as P&G and Unilever have started to sell refills for certain product lines. Which makes sense: I don't need a new spray gun made with unrecyclable plastic every time I buy kitchen cleaner. But any solutions to our waste culture that make life harder for ordinary people is unlikely to ever move beyond a niche.

At times, the new wave of zero-waste consumption can feel like the recycling boom in the 1970s – an earnest popular movement under threat of co-option by the very forces that shunned it. Every time I saw a company make a new zero-waste pledge, I thought of Keep America Beautiful, or Coca-Cola's broken promises of decades past. It was ironic to see: greenwashing tactics being recycled.

For the most part, I began to see much of the new zero-waste lifestyle as what the academic Samantha MacBride calls *busyness*. 'Busy-ness is a handy method of maintaining the status quo yet is simultaneously active, optimistic, and often makes people feel better,' MacBride writes. 'For this reason, to critique its diversionary aspects tends to come across as nay-saying.'[7] Just as corporations pushed recycling and personal responsibility to help defeat legislation that might have harmed their bottom lines, so too the new zero-waste movement is in danger of being co-opted, as it becomes less a policy goal than a branding exercise; a way to sell refillable cups and corn-starch sneakers.

When I first set out to write this book, my plan had been in part to seek out positive stories – the individuals and companies acting to make changes to our wasteful economy. And I saw and met many of them: companies like Apeel Sciences, in

California, which is working to coat fruit and vegetables with a wax-like surface that negates the need for plastic wrapping. Sharing economy apps like Olio and Too Good to Go, which are helping people and businesses cut down on food waste and increase charitable donations. Greyparrot, a British AI company, is using artificial intelligence to increase robotic sorting on recycling lines, to prevent materials being lost to incineration. And numerous companies are pioneering new ways to recycle materials – from textiles to fabrics, from difficult plastics to used electronics. There were (and are!) countless inspirational groups out there, pushing the boundaries of how to reduce, reuse, and recycle our waste. But for every group making an actual difference, there was far more corporate greenwashing by massive multinationals under the guise of the circular economy.

One of the favourite new greenwashing tactics, for example, is for brands to advertise products made of 'ocean plastic' (as opposed to plastic still actually *in* the ocean, which the packaging industry has quietly and deliberately rebranded as 'marine litter'). 'Ocean plastic', contrary to what you might assume, does not mean the plastic has been painstakingly fished out of a Pacific gyre. Rather it refers to plastic intercepted 'within 50km distance of an ocean or major waterway that feeds into the ocean'[8] – a fact far less impressive when you consider that 75 per cent of the world's population live within that area. (A few admirable groups, such as The Ocean Cleanup, *are* actually fishing plastic out of the ocean and rivers, though their efforts to date are tiny compared with the size of the problem.)

Whenever I see big brands discussing the circular economy, I think back to something that Oleksii told me at Pink Recycling. As a textile trader, he'd been invited to events and industry discussions on the circular economy with major retailers. 'When you try to go back to supermarkets and tell them to produce better quality at higher prices, to sell less, they say, "You are talking to the wrong crowd,"' he told me. 'You have to go back to *Capital*, Marx, to understand that what you're trying to do is utopia – it's not gonna work. But if you approach the supermarkets and say, "I can show you how to sell more, make more money." *Then* they will engage with you.' Oleksii painted a picture of a 'circular' economy that is almost the antithesis of its original goal, but which felt all too realistic, if current trends continue. 'You will have a factory that makes seven T-shirts, you buy the T-shirts, wear them – but you don't ever wash them, you just throw them away. Then you take these T-shirts back to the supermarket, and their recycling production line produces out of those T-shirts seven *new* T–shirts – the full loop. And then they sell you the new ones.' Less a closed system of ethical consumption than capitalism as hamster wheel, spinning ever faster. I think about it every time I pass the clothing 'recycling' bins that have recently started appearing outside superstores.

This is the risk that any discussion about the circular econ-omy faces. There is a vision of the circular economy that is about slower consumption, about valuing things, about consuming less and polluting less. But there is also another, more corporate vision, in which circularity simply encourages disposability. This is not hypothetical: a 2013 study, published in the *Journal of Consumer Psychology*, found that if a recycling bin was in the room, participants used twice as much paper (once in an office task, and once in a restroom) as when one wasn't. Other studies have found similar effects.[9] The paper confirmed what Coca-Cola and so many others must, by now, have surmised: the presence of the recycling bin seems to act like a subconscious green light to consume more.[10]

Crucially, I have become convinced that any discussion of circular economies or eradicating waste must confront a simple truth: that while zero waste is at its core an urgent and radical idea with the potential to reshape our world for the better, it is also, in a very real sense, impossible. Recycling eventually comes against physical limits – plastics that can only be recycled a handful of times, pulp fibres that shorten and snap – requiring a constant supply of new feedstock. When corporations show imagery of the 'circular economy', they often show a literal circle, with nothing leaving. Others might show a small stream leaving the loop – like nuclear waste, which is always left behind. However small they draw it, that stream cannot be ignored. The circle will never be a truly closed loop. The real discussion about waste is not about true circularity, but controlling whether that stream becomes a trickle or a torrent. Our job is to ensure as little escapes as we can.

Eventually, the conclusion that I come to is laughably simple: buy less stuff. I recognise that this is not the most original idea, but there's something liberating in it. It feels radical, in a small way, in a world of fast fashion and Amazon Prime, to actively consume less. (It wasn't until after I began my research that I discovered that the three Rs of waste reduction – Reduce, Reuse, Recycle – are not just a catchy slogan, but are actually arranged in order of effectiveness.) It's only when I stop buying things absent-mindedly that I start to suddenly see and feel the relentless bombardment of advertising that I'm exposed to every day – on social media, on TV, going to work. The emails practically scream: 'SALE – 70% off! NEW IN!' Blocking them out is like being cured of tinnitus, a droning I never realised was there.

Buying and living with less is one of the oldest philosophical ideas – asceticism being practised in various forms by ancient religions from Christianity to Buddhism. I don't mean living with less in the minimalist, Marie Kondo, what-sparks-joy sense (Kondo, despite being a big advocate of throwing things away, pays almost zero attention to the fate of those things once you do). Rather, I am interested in creating less waste both downstream of me – what I throw away – and also upstream, in the vast unseen industrial waste created before products ever make it to me.

It's an idea that feels strangely fitting for our present moment. We have so much stuff that sometimes it can feel that we are drowning in it. A 2006 study of homeowners in Los Angeles, California, found that 90 per cent of garage space is now used to store stuff, rather than cars.[11] In the UK, 53 per cent of drivers, myself among them, never put their cars in their garage because they are full of accumulated things,[12] most of which we rarely if ever actually use. I once interviewed a man who ran a house clearance business. As he put it: 'People never die wishing they'd bought more shit.'

Before I set out on my no-shopping spree, Hannah and I set some ground rules: we need to eat, and growing children eventually need new clothes. But for myself, I set out to buy as little non-essential stuff as possible. No new clothes or trainers. No fun-looking kitchen gadgets or video games. (Books are research, and therefore get a pass.) This is not a time-limited trial, I hope, but more the start of a habit of more careful consumption. In practice it also means that I have to take much better care of my things.

I decide to learn to sew, so that when my clothes start to rip (which they do, annoyingly often) I can fix them, instead of buying new ones. So, with patient tutelage from my mother-in-law Joanne, I learn to understand the difference between a running and a blanket stitch, affixing patches to worn elbows and ripped knees. I am, and this is an understatement, not a natural. I often find myself pinpricked and irritable, but it's rewarding work, inducing periods of slow and intense concentration. The threads I use are in bright colours, oranges and yellows, inspired by the Japanese art of *Kintsugi* – in which broken pottery is repaired with gold, to show the beauty of both damage and repair.* I find it strangely moving. I set out to fix things more: broken toys, house fixtures, a coffee machine which has lingered for months with an internal leak that I take to with a screwdriver before seeking professional help, once it's obvious I'm way out of my depth. (Baby steps.) We start to embrace Vinted and other re-use apps, buying the kids' clothes second-hand and selling off the old to new homes. (Hannah has become so adept at this that I feel she missed her natural calling as a trader at Kantamanto.) It's not long before I start to notice a change – and not only in my wallet. In mending and repairing, I'm more connected to what I'm using, and discarding. Suddenly, I see the value in things where once I might not have. What's more, each machine-made thing, torn and fixed, becomes uniquely personal. More human. This is something I'd been told by people in the repair trade: once you learn to fix things, the concept of ownership shifts. What was passive becomes active; the difference is not unlike that between buying cut flowers and learning to tend a garden.

December arrives, and with it the first serious test of my no-new-stuff policy: Christmas. The holiday of giving, and the holiest day on the consumption calendar, a day that dwarfs even Black Friday in terms of sheer wastefulness. The idea of a gift-free Christmas is . . . contentious. Hannah is convinced it will ruin the spirit of the day; to my 4-year-old, the idea of

---

* The Japanese, who it seems are never short of useful terms, have a name for this: *Shashiko*, a kind of running-stitch embroidery used for decorative reinforcement.

a Christmas without presents feels tantamount to child abuse. And so I set out on my own, lonely experiment: no physical presents. (Tickets and vouchers of a non-stuff-purchasing nature, we decide, are acceptable.) For our daughters, we settle on a subscription to Whirli, a toy rental website that lets you send back the toys when kids grow out of them.

When the day itself rolls around I feel, at first, a little strange. Deflated, even, watching other people tear open their gifts out of habit, seeing their joyful faces. But soon I realise that actually, I'm enjoying the day *more*. Without tat to buy, the presents I do get are more thoughtful, more me. There's no awkward pretending to like something that someone else has bought, just to be polite. Without presents as the central focus, we spend more time focused on food and company. And crucially: when the carnage is over and the wrapping is bagged up and thrown away, there's far less waste than ever before. I feel lighter. It's my favourite Christmas in years.

That isn't to say that our lifestyle changes have all been successful, or even particularly effective. We are not, despite everything, zero-waste zealots. There are limits to how far we're willing to go to sacrifice convenience: we microwave rice out of packets too often, and despite my lobbying efforts, we decide to stick with disposable nappies, rather than switching to reusables.[13] I feel guilty about those things, but I also have to remember that some things (convenience, sanitation) are natural human wants and desires. One family sticking to disposable nappies isn't going to destroy the world, just as switching to expensive washable ones is going to save it.

Feeling guilty is also falling into that age-old trap, that personal responsibility is going to be enough to change the system. I wash my recycling fastidiously: that doesn't mean it's not going to end up in a contaminated bale, and spoiled, or shipped to some cement kiln in Indonesia. Does me doing that help to save the planet, or does it help plastic companies avoid a more fundamental change towards (truly) biodegradable materials? Is what I'm doing making a difference, or is it just *busy-ness*? The answer, I have concluded, is something of both.

When I first wrote about Green Recycling and National Sword, a lot of people would say to me: *So, recycling is a myth?* I've spent years researching and thinking about that question, to which my answer is – to my surprise – no. Plastics recycling is deeply flawed, and perhaps irredeemable. But for everything else, recycling is, in my experience, the best option we have to dispose of our unwanted things. We need more recycling, not less. But more fundamentally, we need a waste system that is transparent and, above all, honest. Too much of recycling is built on obfuscation and bullshit. We should eradicate the misleading Resin Codes on plastic packets, and publish actual recycling figures, not those based on guesswork and half-truths. We should stop discussing 'plastic' and instead talk about specific materials – HDPE and PET and LDPE – at school, teaching kids the advantages and limitations of each. Call things by their names. Waste pickers know the value of every material. Once you do, too, you're far less likely to waste it.

Force companies to reveal the actual waste footprint of the goods they're selling – not just in disposal, but from point of extraction. (This practice, also known as the True Cost or True Price movement, is small but growing.) Design out problem materials, such that everything is either recyclable or biodegradable, and ideally both. Make greenwashing illegal, and prosecute it. And we need a fully and equitably funded waste system, one that goes far beyond our broken models of Extended Producer Responsibility, so that those companies producing the overwhelming majority of our waste – industrial, sewage, and household waste – are paying to clean up their waste where it actually ends up, whether that be in the Thames or a Ghanaian gutter.

Things are already moving. Throughout the reporting of this book, global governments have been undertaking the largest shakeup of their waste and recycling laws in decades. Some of the world's most polluting corporations have signed major commitments towards a more circular economy, including Coca-Cola, Unilever, and P&G. In the UK, the Conservative government is planning to introduce deposit return schemes, under a slate of new laws beginning in Scotland in 2023.[14] In the US, federal and state governments have started to get serious(ish) around the Right of Repair. And in 2022, the United Nations agreed in Nairobi to begin consultations on the first ever Global Treaty on Plastic Pollution, which activists hope will include limits on plastic production and stringent new requirements on waste.[15] As I write this conclusion, the first rounds of negotiations are ongoing. It isn't nearly enough, but it's a start.

For my part, my hope is that we stop treating our waste as something secret, to be hidden away. If out of sight is indeed out of mind, then let's put our waste in full view. 'Bring garbage into the open,' Don DeLillo wrote. 'Let people see it and respect it. Don't hide your waste facilities. Make an architecture of waste. Design gorgeous buildings to recycle waste and invite people to collect their own garbage and bring it with them to the press rams and conveyors. Get to know your garbage.' Whether it ends up in the air or the ground, every one of us is constantly creating waste – whether that be our trash, food, clothing, or the raw materials that make them. Much of that waste will outlive us. What will our legacy be?

* * *

One morning towards the end of my journey, I set out from home and drive south, following London's fringes eastwards until I once again find myself on the banks of the Thames, this time at East Tilbury, in Essex. This part of Essex is old waste land: centuries before London began shipping its waste to the rest of the world, it found its dumping grounds here, in the inhospitable flats and marshes of the estuary. Rubbish, plague dead, even the destitute, sent to be rehoused – if London didn't want it, Essex would take it.[16] That status continued well into the twenty-first century. The fittingly named Mucking Marshes, a mile to the west, was once among the biggest land-fills in Europe, taking in 660,000 tonnes a year of London's rubbish until it closed in 2011. It is now (as it was then, in its way) a sanctuary for birds.

I park up at the Coalhouse Fort, a nineteenth-century naval post now used largely by dog walkers, and stride out on the coastal path. It's one of those changeling seaside days, gusty and capricious, heavy clouds gathering on the horizon like youths gearing up for violence. The path is thin and paved, and edged with bushes: blackberries, hawthorn, sea purslane. I follow it south past a radar tower, before dog-legging back upriver towards the city. After half a mile or so, I catch sight of the woman I'm here to meet. Kate Spencer is an environmental geochemist at Queen Mary University of London. A seasoned scientist, Kate arrives looking windswept in a sweater and jeans. She shows me the way down a scruffy bank and onto the beach, where two of her graduate students are setting up transects and making notes. 'Watch your step, it's quite unstable in places,' Kate says. It's low tide, and the beach itself is all rocks and stones. Here, beyond the Thames Barrier, the tide can fluctuate as much as five metres daily, creating a coastline that is in per-ennial flux. A stripe of seaweed marks high water. A warning.

We walk up to where the dunes give way to the beach. There, in the crumbling sands, I see it: layers upon layers of waste. Ribbons of bright fabric. Dull strips of plastic. Brown bottles, green bottles, fully intact. Shards of old china. Newspapers collapsing at the edges, their ink still startlingly clear. This entire beach is holding back a historic landfill; a dump submerged in the sand.

'I don't know if you've heard of the Anthropocene,' Kate says. 'It's a question. The question is: have we created a stratigraphic marker of human presence?' She gestures to the waste, which lies thick like a sandwich filling in the sand. 'I think that's it. You can see it.'

There are approximately 21,000 historic landfills in England and Wales, according to the UK's Environment Agency.[17] Of those, more than 1,000 are known to contain hazardous waste. Several thousand more predate detailed data collection, meaning their exact contents are unknown.[18] According to Kate's research, more than 1,200 historic landfills lie along Britain's coasts and rivers. With sea levels rising and extreme weather events more frequent, these landfills are at high risk of erosion. Some of them, as here at East Tilbury, are already collapsing into the sea. This is not just a British problem: in Europe alone, there are estimated to be as many as 500,000 historic landfills, an untold number of those on coastal floodplains. In the US, more than 900 Superfund sites (including Tar Creek) are under threat of breaching as sea levels rise.[19]

'It's a pollution source that nobody thought about ten years ago,' Kate says. Historic landfills can still be rich with contaminants: lead, asbestos, plastics. In 2016, Kate published a study conducted alongside the Environment Agency, which found elevated levels of heavy metals including lead along this beachline, along with polycyclic aromatic hydrocarbons (PAHs), a toxic carcinogen. 'We know things like arsenic can be very elevated,' Kate says. Other studies have found PCBs leaching from similar sites; it is thought that such slow releases may bio-accumulate and amplify up the food chain, into animals such as whales. In 2017, an orca was found off the coast of the UK which contained in its flesh the largest concentrations of PCBs ever recorded in animals.[20] Slow releases of persistent chemicals from landfills are one suspect.

We wander along the beach, mudlarking – looking for treasures among the trash. A blue china plate in fragments reminds me of my mother-in-law. There are tiny black lead batteries, their wiring gone. 'This is a pair of tights,' Kate says, pulling some nylons from the sand face. She tugs at them: still stretchy. 'These are nylons and polyesters. They're essentially plastics. They're still here after seventy years, and haven't degraded. They're full of colour.' You can almost make out decades in the strata, the brown geometric Sixties prints giving way to lush Seventies-style paisleys. What was it DeLillo wrote? *Waste is the secret history, the underhistory*. Everything we bury is a time capsule, its fate unknown. The problem is that waste has a habit of resurfacing.

I spot some fragments of old newspaper, the pages stuck together by salt and moisture. The dates are gone, but there's a mention of David Lloyd George, the UK prime minister from 1916 to 1922. Kate herself has found intact fragments dating to at least the 1930s. I pick up one page, with a cartoon half obscured. All I can make out is the caption. It says: THE AFTERMATH.

'Our production of solid waste is increasing at a greater rate than any other pollutant,' Kate says. 'People think it's a problem that we've dealt with. They think, *We're much better now, we recycle now, we don't landfill any more*.' She picks out a packet of something, the text long-since faded. 'This is smacking you in the face, that we can't forget about it. We can't forget that all of this historic pollution that we've caused is just . . . buried.'

When Kate first started releasing her studies, the waste industry reacted with denial. 'People would say, "Well, it's been buried fifty years, it's all degraded, it's all gone away." And actually, the work we did we found: well, no, it's still releasing ammonia. These sites are still releasing things,' she says. The discovery has upended assumptions about how long it can take waste inside landfills to decay. 'In a way, it [landfill] is a really long experiment that we haven't got to the end of yet.'

I scoop up some of the fragments into my hands: crockery shards, an old lead battery, a plastic Spirograph, its whorl like a Nautilus shell. It's a funny thing, permanence. Human beings spend so much thought and care and energy on making things, and throw them away without a thought. But they don't disappear. 'The waste remains,' as the poet William Empson wrote. Whether we bury it in the ground, or burn it, releasing the carbon within into the air. Matter is matter is matter. 'We don't call it waste,' Jamie told me, all the way back at the beginning of my journey, at Green Recycling. 'We call it materials.' From a geological perspective, it's all the same.

After a while longer I say my goodbyes, leaving the scientists to their study, and climb back onto the coastal path. The sun has broken through, loosing shards of light onto the surface of the estuary. A little way down the track, I spot some rubbish: a Lucozade bottle discarded in the scrub. I pick the bottle up, rub my fingers over its plastic sleeve (PET again), feel the serrated lid under my thumb. Who made this thing? How many lives has it touched, whose hands, whose lips? How many decades or centuries would it persist on this Earth, should I just leave it here, as someone else has? These are the things I think about idly as I walk back to my car. There is a story in everything we throw away, if only we choose to look.

By the time I get home I've forgotten about the Lucozade bottle. Instead it languishes in the passenger-side footwell until a couple of days later, when I finally remember to crush the bottle down, screw the lid back on, and toss it in the recycling bin, to continue its journey.

ACKNOWLEDGEMENTS

This journey began in 2019, when I sent my editor at *The Guardian* a speculative email: 'Would you be at all interested in a feature about recycling? (It's more interesting than it sounds, honest.)' Thank you to Rob Fearn and the team at what is now the *Guardian Saturday* magazine for saying yes when few would have. I did not expect then that that email would be the start of a four-year odyssey. In the time since, my eldest daughter has gone from a baby to starting school; we've lived through a global pandemic; I developed a chronic health condition; and we've had another new addition to the family. Throughout I've been lucky to have the support of many people, without whom this book would not be in your hands.

My deepest thanks to my agent, Chris Wellbelove, who believed in me and my little rubbish book since the beginning. Huge thanks also to my editors at Simon & Schuster, Holly Harris, Kat Ailes, Alex Eccles and Frances Jessop, and to Sam Raim and Lauren Marino at Hachette in New York, who stuck by me even when the pandemic grounded my reporting for more than a year. Hayley Campbell, Madhu Murgia and Daniella Graham read early drafts and provided innumerable suggestions, for which I am truly grateful.

This book relied on interviews with many dozens of people, all over the world. Many of them are mentioned in these pages, but far more didn't make the final draft. Their insights and wisdom have nonetheless been crucial in my journey and shaping my thinking about waste. Among them, I'd like to thank Jim Puckett and his team at the Basel Action Network; Jan Dell, of the Last Beach Cleanup; the GAIA network; Robin Ingenthron; Joshua Goldstein; Sedat Gundogdu; Theo Thomas, Janet Gunter and Ugo Valleri of the Restart Project; Roland Geyer; Virginia Comolli; Kyle Wiens; and everyone at WIEGO.

In India, I was grateful to work with Rahul Singh and Yashraj Sharma, who provided fixing and translation services in Kanpur and New Delhi, respectively. Both are able reporters in their own right, and contributed so much to my reporting. Thanks also to Sheikh Akbar Ali, and the teams at WIEGO, Basti Suraksha Manch, and Toxics Link; Prof. Vishwambhar Nash Mishra; and the researchers at IIT Kanpur.

In Ghana, I was aided by my able guide and fixer, Sena Kpodo. Thanks again to The Revival, to everyone at the OR Foundation, and the many traders of Kantamanto. In Malaysia, my thanks to Mageswari Sangaralingam, Prigi Arisandi, and Yeo Bee Yin. Steve Wong and I have still never met in person, but his insights into the global export markets and to China's recycling history proved invaluable. The pandemic prevented me from fulfilling my dream of visiting Onkalo, but thanks to Pasi Tuohimaa and the team there, I feel like I already have.

In Oklahoma, my thanks to Summer King, Craig Kreman and everyone at the Quapaw Nation Environmental Office; Rebecca Jim and her team at Lead Agency; and to Ed Keheley, for his invaluable history of Tar Creek and the Superfund site. Thanks also to John Sherigian and everyone at Electronics Recycling International in Fresno.

In the UK, I have met and been helped by far too many people to name. I am particularly indebted to Adam Read and

Acknowledgements

his team at SUEZ; to Trevor Bradley, and the wonderful gleaners of Feedback; to Tim Price and the communications team at DS Smith; to Krista Lord at Biffa Polymers; to Ben Chilton at Sellafield; to Oleksii Kotyk; and to the team at Thames Water. Danielle Purkis and Mark Miodownik were extremely helpful at a busy time, and their research into compostables and other materials is truly exciting. An especially wholehearted thanks to Compost John Cossham; and to Jamie Smith and Green Recycling, who started it all.

Writers don't often thank other writers, but journalism of this kind leans so much on the dogged reporting of others. In the case of waste, I'd be remiss without highlighting a few: in particular Sharon Lerner of the *Intercept*, Rachel Salvidge and Sandra Laville at *The Guardian*, the waste historian Rebecca Altman, and Adam Minter, whose work on Kantamanto in particular helped to shape my thinking about colonialism and waste.

Finally, to my family: my mum and my mother-in-law Joanne, both zero-waste influencers in their own small ways. Hannah . . . thank you. I love you. Matilda, Clemency: you might not understand this until you're older, but this planet is a precious and remarkable thing. Try not to waste it.

NOTES

INTRODUCTION: THE TIPPING FLOOR

1   Imogen Napper, Bede F. R. Davies et al., 'Reaching New Heights in Plastic
    Pollution – Preliminary Findings of Microplastics on Mount Everest', *One
    Earth* 3(5), 2020: DOI: 10.1016/j.oneear.2020.10.020
2   A.J. Jamieson, L. Brooks et al. 'Microplastics and synthetic particles
    ingested by deep-sea amphipods in six of the deepest marine ecosystems on
    Earth', *Royal Society Open Science* 6(3), 2019: DOI: 10.1098/rsos.180667
3   United Nations Environment Programme, *From Pollution to Solution: A Global
    Assessment of Marine Litter and Plastic Pollution*, 2021.
4   Plastic Pollution Coalition, 'New Research Shows The Great Pacific
    Garbage Patch is 3 Times The Size of France', 2018: https://www.
    plasticpollutioncoalition.org/blog/2018/3/23/new-research-shows-the-
    great-pacific-garbage-patch-is-3-times-the-size-of-france
5   Raffi Khatchadourian, 'The Elusive Peril Of Space Junk', *New Yorker*,
    21/09/2020: https://www.newyorker.com/magazine/2020/09/28/
    the-elusive-peril-of-space-junk
6   S. Kaza, L. Yao, P. Bhada-Tata and F. Van Woerden, 'What A Waste 2.0:
    A Global Snapshot of Solid Waste Management to 2050', The World Bank
    (2018): DOI: 10.1596/978-1-4648-1329-0
7   Ibid.
8   Sandra Laville and Matthew Taylor, 'A million bottles a minute: world's
    plastic binge "as dangerous as climate change"', *The Guardian*, 28/06/2017:
    https://www.theguardian.com/environment/2017/jun/28/a-million-a-
    minute-worlds-plastic-bottle-binge-as-dangerous-as-climate-change
9   Tik Root, 'Cigarette butts are toxic plastic pollution. Should
    they be banned?', *National Geographic*, 09/08/2019: https://www.
    nationalgeographic.com/environment/article/cigarettes-story-of-plastic
10 'UK Statistics on Waste', UK Department for Environment, Food &
    Rural Affairs, 11/05/2022 (accessed: 22/06/2022): https://www.gov.uk/
    government/statistics/uk-waste-data/uk-statistics-on-waste

11   Kaza et al, 'What A Waste 2.o', p. 18.
12   Nick Squires, 'Mafia accused of sinking ship full of radioactive waste off
       Italy', *The Telegraph*, 16/09/2009: https://www.telegraph.co.uk/news/
       worldnews/europe/italy/6198228/Mafia-accused-of-sinking-ship-full-of-
       radioactive-waste-off-Italy.html
13   Jandira Morais, Glen Corder et al., 'Global review of human waste-
       picking and its contribution to poverty alleviation and a circular economy',
       Environmental Research Letters, 2022. See also International Labour
       Organization, *Sustainable development, decent work and green jobs* (2013).
14   Martin V. Melosi, *Garbage in the Cities* (Pittsburgh: University of Pittsburgh
       Press), revised edition, 2004, p. 4.
15   Charles Dickens, *Bleak House* (London: Penguin), 2003, p. 60.
16   Janice Brahney, Margaret Hallerud et al., 'Plastic rain in protected areas of
       the United States', *Science* (2020): DOI: 10.1126/science.aaz5819
17   Adam Minter, *Junkyard Planet: Travels in the Billion-Dollar Trash Trade* (New
       York: Bloomsbury Press), 2013, p. 12.

1. THE MOUNTAIN

1   The other two, Bhanpur and Bhalswa, are nearly as large.
2   Jandira Morais, Glen Corder et al., 'Global review of human waste-
      picking and its contribution to poverty alleviation and a circular economy',
      Environmental Research Letters (2022). See also International Labour
      Organization, *Sustainable development, decent work and green jobs* (2013).
3   Due to the irregularity of the census and the dense population of Delhi's
      informal settlements, or *jhuggis*, this may be an undercount.
4   S. Kaza, L. Yao, P. Bhada-Tata and F. Van Woerden, 'What A Waste 2.o:
      A Global Snapshot of Solid Waste Management to 2050', The World Bank
      (2018): DOI: 10.1596/978-1-4648-1329-0
5   Romie Stott, 'Oxyrhynchus, Ancient Egypt's Most Literate Trash Heap',
      Atlas Obscura, 16/03/2016: https://www.atlasobscura.com/articles/
      oxyrhynchus-ancient-egypts-most-literate-trash-heap
6   https://indianexpress.com/article/cities/delhi/ghazipur-landfill-garbage-
      mound-collapse-kills-2-4824659/
7   Himal Kotelawala, 'Sri Lanka Death Toll Rises in Garbage Dump
      Collapse', *New York Times*, 17/04/2017: https://www.nytimes.
      com/2017/04/17/world/asia/sri-lanka-garbage-dump.html
8   PET – Polyethylene terephthalate; HDPE – high-density polyethylene;
      PVC – polyvinyl chloride; PP – polypropylene.
9   This is the low end. Within a twenty-year time window, methane is eighty-
      four times more warming than CO_2, according to the IPCC: https://www.
      factcheck.org/2018/09/how-potent-is-methane/

10 'Methane Menace: Aerial survey spots "super-emitter" landfills', Reuters, 18/06/2021: https://www.reuters.com/business/sustainable-business/methane-menace-aerial-survey-spots-super-emitter-landfills-2021-06-18

11 Will Mathis and Akshat Rathi, 'Methane Plumes Put Pakistan Landfills In The Spotlight', *Bloomberg*, 30/09/2021: https://www.bloomberg.com/news/articles/2021-09-20/methane-plumes-in-pakistan-put-landfills-in-the-spotlight

12 E. R. Rogers, R. S. Zalesny and C. H. Lin, 'A systematic approach for prioritizing landfill pollutants based on toxicity: Applications and opportunities', *Journal of Environmental Management* 284 (2021).

13 Superfund National Priorities List, US Environmental Protection Agency, 2022: https://www.epa.gov/superfund/national-priorities-list-npl-sites-site-name

14 It's worth noting that an association does not prove causality – but also that establishing causality, particularly when it comes to cancer, is devilishly difficult to establish with any kind of chemical or environmental hazard, largely due to the number of confounding factors and timescales involved. In the Global South, living near landfills is also associated with higher rates of cholera, diarrhoea, malaria, and tuberculosis.

15 William Rathje and Cullen Murphy, *Rubbish! The Archaeology of Garbage*, (New York: HarperCollins), 1992.

16 Elizabeth Royte, *Garbage Land: On the Secret Trail of Trash*, (New York: Back Bay Books), 2005, p. 60.

17 US Bureau Of Labor Statistics, 2020.

18 United States Environmental Protection Agency, 'National Overview: Facts on Materials, Wastes and Recycling', 2022: https://www.epa.gov/facts-and-figures-about-materials-waste-and-recycling/national-overview-facts-and-figures-materials; Australian Bureau of Statistics, 'Waste Account, Australia, Experimental Estimates: 2018–2019 financial year', 2020: https://www.abs.gov.au/statistics/environment/environmental-management/waste-account-australia-experimental-estimates/latest-release

2. SAVE SCRAP FOR VICTORY!

1 'Wrapping up? How paper and board are back on track', *Allianz Trade*, 2021: https://www.allianz-trade.com/en_global/news-insights/economic-insights/wrapping-up-how-paper-and-board-are-back-on-track.html

2 World Wildlife Fund, *WWF Living Forests Report: Chapter 4 – Forests and Wood Products*, 2012, p. 10.

3 US Environmental Protection Agency, 'Advancing Sustainable Materials Management: 2018 Fact Sheet', December 2020: https://www.epa.gov/

sites/default/files/2021-01/documents/2018_ff_fact_sheet_dec_2020_
fnl_508.pdf

4   Konrad Olejnik, 'Water Consumption in Paper Industry – Reduction
    Capabilities and the Consequences' (2011): DOI: 10.1007/978-94-007-
    1805-0_8

5   Daniel Thomas and Jonathan Eley, 'UK businesses suffer cardboard
    shortages due to "Amazon effect"', *Financial Times,* 12/03/21: https://www.
    ft.com/content/21b735db-4827-4aed-a16e-74aaa7c4c67b

6   Chris Stokel-Walker, 'Amazon is hoarding all the boxes. That's bad
    news for eggs', *WIRED,* 17/02/2021: https://www.wired.co.uk/article/
    amazon-cardboard-boxes-recycling

7   Interestingly, the researchers suggested that besides practicality, another
    reason to reuse old tools might have been to honour the ancestors who
    made them. Bar Efrati et al., 'Function, life histories, and biographies
    of Lower Paleolithic patinated flint tools from Late Acheulian Revadim,
    Israel', *Scientific Reports,* 2022: DOI: 10.1038/s41598-022-06823-2

8   Dennis Duncan, 'Pulp non-fiction', *Times Literary Supplement,* 30/09/2016.

9   Jessica Leigh Hester, 'The Surprising Practice of Binding Old Books With
    Scraps of Even Older Books', *Atlas Obscura,* 11/6/2018: https://www.
    atlasobscura.com/articles/book-waste-printed-garbage

10 Henry Mayhew, *London Labour and the London Poor* (Oxford: Oxford
    University Press), 2010, p. 162.

11 Ibid., p. 149.

12 Ibid., p. 178.

13 Melosi, *Garbage in the Cities*, p. 35.

14 S. E. Finer, *The Life and Times of Sir Edwin Chadwick* (London: Taylor &
    Francis), 1952, p. 243.

15 Edwin Chadwick, *Report on the Sanitary Condition of the Labouring Population
    of Great Britain*, 1842, p. 6. Accessed via wellcomecollection.org.

16 One upside of the miasma theory is that Victorians were obsessed with
    ventilation, which may have helped fight the spread of airborne viruses –
    and gained a renewed appreciation during the Covid-19 pandemic.

17 Melosi, *Garbage in the Cities*, p. 48.

18 Heather Rogers, *Gone Tomorrow: The Hidden Life of Garbage* (New York: The
    New Press), 2005, p. 54.

19 Stefano Capuzzi and Giulio Timelli, 'Preparation and Melting of Scrap in
    Aluminium Recycling: A Review', *Metals* (2018): DOI: 10.3390/met8040249

20 *Benefits of Recycling*, Stanford University, https://lbre.stanford.edu/
    pssistanford-recycling/frequently-asked-questions/frequently-asked-
    questions-benefits-recycling

21 US Environmental Protection Agency, 'Environmental factoids', 30/03/2016:
    https://archive.epa.gov/epawaste/conserve/smm/wastewise/web/html/
    factoid.html

22   Kirsten Linninkoper, 'Multi-billion growth ahead for international scrap recycling market', *Recycling International*, 18/02/2019: https://recyclinginternational.com/research/multi-billion-growth-ahead-for-international-scrap-recycling-market/18639/

23   According to a meta-analysis conducted by GAIA; previous studies have estimated smaller differences in job creation, but still put recycling at least 10 times higher (see Samantha MacBride, *Recycling Reconsidered* (Cambridge, MA: MIT Press), 2013). J. Ribeiro-Broomhead, and N. Tangri, 'Zero Waste and Economic Recovery: The Job Creation Potential of Zero Waste Solutions', Global Alliance for Incinerator Alternatives, 2021: DOI: 10.46556/GFWE6885

24   Bureau of International Recycling, 'World Steel Recycling in Figures: 2015–2019', 2020, p. 2.

25   Zachary Skidmore, 'The fragmentation of the copper supply chain', *Mine* (2022): https://mine.nridigital.com/mine_may22/fragmentation_copper_supply_chain

26   Susan Freinkel, *Plastic: A Toxic Love Story* (New York: Houghton Mifflin Harcourt), 2011, p. 60.

27   Ibid, p. 25

28   Rogers, *Gone Tomorrow*, p. 114.

29   In truth, bottle manufacturers had attempted to patent ways to stop reuse for decades. See Rogers and Strasser for more.

30   Susan Strasser, *Waste and Want* (New York: Henry Holt and Company), 1999, p. 271.

31   Some bacteria have been discovered that can consume some plastics, although research is at an early stage.

32   A. Isobe, T. Azuma, M. R. Cordova et al., 'A multilevel dataset of microplastic abundance in the world's upper ocean and the Laurentian Great Lakes', *Microplastics & Nanoplastics* (2021): DOI: 10.1186/s43591-021-00013-z

33   'Biffa Opens Plastic Recycling Plant in Seaham', Biffa.co.uk, 29/01/20: https://www.biffa.co.uk/media-centre/news/2020/biffa-opens-plastic-recycling-plant-in-seaham

34   Sandra Laville and Matthew Taylor, 'A million bottles a minute: world's plastic binge "as dangerous as climate change"', *The Guardian*, 28/06/2017.

35   Jim Cornall, 'PACCOR to include minimum 30% rHDPE content in UK dairy caps', *Dairy Reporter*, 21/02/2022.

36   Originally the 'voluntary plastic container coding system'.

37   Cooper, who cited the infinite artworks of M. C. Escher as his inspiration, won $2,500 for his effort.

38   Broadly speaking, the numbers indicate (take a deep breath):
1: Polyethylene terephthalate (PET), used in soda bottles and clothing;
2: high-density polyethylene (HDPE), used in milk and juice bottles;

3: polyvinyl chloride (PVC), used in raincoats, pipework and wires; 4: low-density polyethylene (LDPE) used in films, bags and wraps; 5: polypropylene (PP), popular for tubs and yogurt pots; 6: polystyrene (PS), which comes in many forms, from the clear hard plastic used for food containers to the expanded foams used in packaging and take-away boxes; finally, 7: which means 'other', and covers all plastics not included in the first six, including polycarbonate, acrylic, acrylonitrile butadiene styrene (ABS), the plastic used for LEGO bricks, and plant-based plastics such as polylactic acid – none of which are widely recycled, if at all.

39 Laura Sullivan, 'How Big Oil Misled The Public Into Believing Plastic Would Be Recycled', *PBS Frontline*, and NPR, 11/09/2020: https://www. npr.org/2020/09/11/897692090/how-big-oil-misled-the-public-into-believing-plastic-would-be-recycled

40 Ibid.

41 The industry's recent sop to this has been to change the three arrows for a simpler triangle, although at the time of writing that proposal still hasn't been widely adopted.

42 WRAP, 'On-pack Labelling and Citizen Recycling Behaviour', 2020: https://wrap.org.uk/sites/default/files/2021-09/WRAP-On-pack-labelling-and-recycling-behaviour_0.pdf

43 Strasser, *Waste and Want*, p. 261.

44 Emily Cockayne, *Rummage* (London: Profile Books), 2020, pp. 58–60.

45 Rogers, *Gone Tomorrow*, p. 142.

46 Frank Trentmann, *Empire of Things: How We Became a World of Consumers, from the Fifteenth Century to the Twenty-First* (London: Penguin), 2016, p. 639.

47 Mark Kaufman, 'The carbon footprint sham', *Mashable*, 13/07/2020: https://mashable.com/feature/carbon-footprint-pr-campaign-sham

48 'Packaging lobby's support for anti-litter groups deflects tougher solutions', *Corporate Europe*, 28/03/2018: https://corporateeurope.org/en/power-lobbies/2018/03/packaging-lobby-support-anti-litter-groups-deflects-tougher-solutions

49 Myra Klockenbrink, 'Plastic Industry, Under Pressure, Begins to Invest in Recycling', *New York Times*, 30/08/1988: https://www.nytimes.com/1988/08/30/science/plastics-industry-under-pressure-begins-to-invest-in-recycling.html

50 Sullivan, 'How Big Oil Misled The Public Into Believing Plastic Would Be Recycled'.

51 Ibid.

52 Maeve McClenaghan, 'Investigation: Coca-Cola and the "fight back" against plans to tackle plastic waste', *Greenpeace Unearthed*, 25/01/2017: https://unearthed.greenpeace.org/2017/01/25/investigation-coca-cola-fight-back-plans-tackle-plastic-waste/

53 *Branded, Volume IV: Holding Corporations Accountable for the Plastic &*

Climate Crisis, Break Free From Plastic, 25/10/2021: https://www.breakfreefromplastic.org/2021/10/25/the-coca-cola-company-and-pepsico-named-top-plastic-polluters-for-the-fourth-year-in-a-row/

54 Rogers, *Gone Tomorrow*, p. 215.

55 'Coca-Cola Sets Goal to Recycle or Reuse 100 Percent of Its Plastic Bottles in the US', 05/09/2007: https://investors.coca-colacompany.com/news-events/press-releases/detail/274/coca-cola-sets-goal-to-recycle-or-reuse-100-percent-of-its

56 Judith Evans, 'Coca-Cola and rivals fail to meet plastic pledges', 16/09/2020: https://www.ft.com/content/bb189a2a-57ca-44ce-82ab-1d015a20ca1c

57 Sharon Lerner, 'Waste Only', *The Intercept*, 20/07/19: https://theintercept.com/2019/07/20/plastics-industry-plastic-recycling/

58 Tik Root, 'Inside the long war to protect plastic', *Center for Public Integrity*, 16/05/2019: https://publicintegrity.org/environment/pollution/pushing-plastic/inside-the-long-war-to-protect-plastic/

59 'Preemption laws', Plasticbaglaws.org, 2010: https://www.plasticbaglaws.org/preemption/

60 Helene Wiesinger, Zhanyun Wang and Stefanie Hellweg, 'Deep Dive into Plastic Monomers, Additives, and Processing Aids', *Environmental Science & Technology* (2021): DOI: 10.1021/acs.est.1c00976

61 Patricia L. Corcoran, Charles J. Moore and Kelly Jazvac, 'An anthropogenic marker horizon in the future rock record', *GSA Today* (2014): DOI: 0.1130/GSAT-G198A.1

62 Karen McVeigh, 'Nurdles: the worst toxic waste you've probably never heard of', *The Guardian*, 29/11/21: https://www.theguardian.com/environment/2021/nov/29/nurdles-plastic-pellets-environmental-ocean-spills-toxic-waste-not-classified-hazardous

63 This is the EPA's estimate, although the metric varies on the quality and type of paper used; DS Smith says it's twenty-five times. Either way, the point is that fibre is not a fully circular material.

64 Another tactic is to simply add virgin plastic during the liquid phase, along with masterbatch.

65 Spyridoula Gerassimidou, Paulina Lanska et al,. 'Unpacking the complexity of the PET drink bottles value chain: A chemicals perspective', *Journal of Hazardous Materials*, 15/05/22: DOI: 10.1016/j.jhazmat.2022.128410

66 David Shukman, 'Plastic particles found in bottled water', BBC News, 15/03/18: https://www.bbc.co.uk/news/science-environment-43388870

67 Yufei Wang and Haifeng Qian, 'Phthalates and Their Impacts on Human Health', *Healthcare* (2021): DOI: 10.3390/healthcare9050603

68 An endocrine-disrupting chemical, exposure to BPA has been linked to cancer, obesity, and developmental problems, although neither the UK nor US governments have found enough evidence to ban its use.

69 The government quietly missed its target to recycle 50 per cent of

household waste by 2020, though since Brexit it is no longer bound by many EU waste rules.

70　When I found this out, I was incredulous. So I contacted the Department for Environment, Food & Rural Affairs, which controls recycling policy. A press officer replied, 'I can confirm that your understanding is correct.'

71　The document is 'Local Authority Waste statistics – Recycling measures' (2020): https://assets.publishing.service.gov.uk/government/uploads/system/uploads/attachment_data/file/966604/Recycling_Explainer_FINAL_3_accessible.pdf

72　To understand the miscalculation, it's worth noting that in 2008 the EU passed a Waste Framework Directive that bound every EU member to hit 50 per cent household recycling by 2020. The UK was not the only country to miss that target.

73　Alex Gray, 'Germany recycles more than any other country', weforum.org, 18/12/2017: https://www.weforum.org/agenda/2017/12/germany-recycles-more-than-any-other-country/

74　The Last Beach Cleanup & Beyond Plastics, 'The Real Truth About the U.S. Plastic Recycling Rate: 2021 U.S. Facts And Figures', 04/05/2022: https://static1.squarespace.com/static/5eda91260bbb7e7a4bf528d8/t/6 2726edceb7cc742d53eb073/1651666652743/The+Real+Truth+about+ the+U.S.+Plastic+Recycling+Rate+2021+Facts+and+Figures+_5-4- 22.pdf

75　'Chemical Recycling of Sachet Waste: A Failed Experiment', GAIA, January 2022: https://www.no-burn.org/unilever-creasolv/

76　Joe Brock, Valerie Volcovici and John Geddie, 'The Recycling Myth: Big Oil's Solution For Plastic Waste Littered With Failure', Reuters, 29/07/21: https://www.reuters.com/investigates/special-report/environment-plastic-oil-recycling/

77　Kit Chellel and Wojciech Moskwa, 'A Plastic Bag's 2,000-Mile Journey Shows the Messy Truth About Recycling', *Bloomberg Businessweek*, 29/03/2022: https://www.bloomberg.com/graphics/2022-tesco-recycle-plastic-waste-pledge-falls-short

3. THE WORLD'S GARBAGE CAN

1　'Discarded: Communities on the Frontlines of the Global Plastic Crisis', GAIA, April 2019: https://www.no-burn.org/resources/discarded-communities-on-the-frontlines-of-the-global-plastic-crisis/

2　Alice Ross, 'UK household plastics found in illegal dumps in Malaysia', *Greenpeace Unearthed*, 21/10/2018: https://unearthed.greenpeace.org/2018/10/21/uk-household-plastics-found-in-illegal-dumps-in-malaysia/

3   'TRASHED: How the UK is still dumping plastic waste on the rest of the
    world', Greenpeace, 17/05/2021: https://www.greenpeace.org.uk/wp-
    content/uploads/2021/05/EMBARGOED-GPUK-Trashed-report.pdf

4   Alan Cox, Peter Hounsell, Sue Kempsey et al., 'The King's Cross Dust
    Mountain and the Bricks to Rebuild Moscow after 1812' (London: British
    Brick Society), 2017: http://britishbricksoc.co.uk/wp-content/uploads/
    2018/05/BBS_137_2017_Nov.pdf

5   Adam Minter, *Secondhand: Travels in the New Global Garage Sale* (New York:
    Bloomsbury), 2019, p. 117.

6   Joshua Goldstein, *Remains of the Everyday: A Century of Recycling in Beijing*
    (Berkeley: University of California Press), 2020, p. 47.

7   Will Flower, 'What Operation Green Fence has Meant for Recycling',
    Waste 360, 11/02/2016: https://www.waste360.com/business/what-
    operation-green-fence-has-meant-recycling

8   Minter, *Junkyard Planet*, p. 7.

9   Goldstein, *Remains of the Everyday*, p. 278.

10 Minter, *Junkyard Planet*, p. 201; Jack Hunter, '"Flood" of toxic Chinese
    toys threatens children's health', European Environmental Bureau, 2019:
    https://eeb.org/flood-of-toxic-chinese-toys-threatens-childrens-health.

11 Goldstein, *Remains of the Everyday*, p. 357.

12 Yining Zou, 'Where Do the Plastic Miners Go When the "Mine"
    Disappears?', *New Security Beat*, 16/01/2020: https://www.newsecuritybeat.
    org/2020/01/plastic-miners-mine-disappears/

13 Minter, *Junkyard Planet*, p. 145.

14 Goldstein, *Remains of the Everyday*, p. 351.

15 Ibid., p. 379.

16 Li Heng, '"Recycling economy" brings wealth, pollution', *China News
    Service*, 7/05/2011: http://www.ecns.cn/in-depth/2011/07-05/444.shtml

17 Anna Leung, Zong Wei Cai and Ming Hung Wong, 'Environmental
    contamination from electronic waste recycling at Guiyu, southeast China',
    Journal of Material Cycles and Waste Management (2006): DOI: 10.1007/
    s10163-006-0002-y

18 Ministry of Environmental Protection of the People's Republic of China
    to World Trade Organization, 18/07/2017: https://docs.wto.org/
    dol2fe/Pages/SS/directdoc.aspx?filename=q:/G/TBTN17/CHN1211.
    pdf&Open=True

19 Kiki Zhao, 'China's Environmental Woes, in Films That Go Viral, Then
    Vanish', *New York Times*, 28/04/2017.

20 European Commission, *A European Strategy for Plastics in a Circular Economy*,
    2018, p. 16.

21 Amy L. Brooks, Shunli Wang and Jenna Jambeck, 'The Chinese import
    ban and its impact on global plastic waste trade', *Science Advances*, 20/06/18:
    DOI: 10.1126/sciadv.aat0131

22　Colin Staub, 'China: Plastic imports down 99 percent, paper down a third', *Resource Recycling*, 29/01/2019: https://resource-recycling.com/recycling/2019/01/29/china-plastic-imports-down-99-percent-paper-down-a-third/

23　Associated Press, 'California's largest recycling center business, RePlanet, shuts down', *Los Angeles Times*, 05/08/2019: https://www.latimes.com/business/story/2019-08-05/recycling-center-business-replanet-shuts-down

24　The Heard and McDonalds Islands received 57 tonnes of unsorted waste from the US in 2016 – probably an attempt by waste traders to mask shipments' origin and contents from their eventual port of destination. Xavier A. Cronin, 'America's plastic scrap draft', *Recycling Today*, 30/09/2016.

25　Karen McVeigh, 'Huge rise in US plastic waste shipments to poor countries following China ban', *The Guardian*, 05/10/2018: https://www.theguardian.com/global-development/2018/oct/05/huge-rise-us-plastic-waste-shipments-to-poor-countries-china-ban-thailand-malaysia-vietnam

26　Richard C. Paddock, 'To Make This Tofu, Start by Burning Toxic Plastic', *New York Times*, 14/11/19: https://www.nytimes.com/2019/11/14/world/asia/indonesia-tofu-dioxin-plastic.html

27　Greenpeace, 'Malaysia endures toxic legacy of UK plastic waste exports', Greenpeace, 27/05/20: https://www.greenpeace.org.uk/news/malaysia-endures-toxic-legacy-of-uk-plastic-waste-exports

28　Michelle Tsai, 'Why the Mafia Loves Garbage', *Slate*, 11/01/2008.

29　Heather Rogers, *Gone Tomorrow: The Hidden Life of Garbage* (New York: The New Press), 2005, p. 189.

30　There is even more detail of New York's fight with the garbage gangs in Rick Cowan and Douglas Century, *Takedown: The Fall of the Last Mafia Empire* (New York: GP Putnam's Sons), 2002.

31　Jake Adelstein, 'Why One Of Japan's Largest Organized Crime Groups Is Looking For Legitimate Work', *Forbes*, 02/10/2017.

32　Juan José Martínez D'Aubuisson, 'How the MS13 Became Lords of the Trash Dump in Honduras', *InSight Crime*, 19/01/22: https://insightcrime.org/news/honduras-how-ms13-became-lords-trash-dump/

33　Virginia Comolli, *Plastic For Profit: Tracing illicit plastic waste flows, supply chains and actors* (Geneva: Global Initiative Against Transnational Organized Crime), 2021, p. 4.

34　Jim Yardley, 'A Mafia Legacy Taints the Earth in Southern Italy', *New York Times*, 29/01/2014: https://www.nytimes.com/2014/01/30/world/europe/beneath-southern-italy-a-deadly-mob-legacy.html

35　NAO, 'Investigation into government's actions to combat waste crime in England', National Audit Office, 27/04/2022: https://www.nao.org.uk/press-release/investigation-into-governments-actions-to-combat-waste-crime-in-england

36 Louise Smith, 'Research Briefing: Fly-tipping – the illegal dumping of waste', UK House Of Commons (London: Commons Library), 23/05/2022: https://commonslibrary.parliament.uk/research-briefings/sn05672

37 NAO, 'Investigation into government's actions to combat waste crime in England'.

38 Comolli, *Plastic For Profit*, p. 33.

39 Rogers, *Gone Tomorrow*, p. 195.

40 *Securities and Exchange Commission v. Dean L. Buntrock, Phillip B. Rooney, James E. Koenig, Thomas C. Hau, Herbert A. Getz, and Bruce D. Tobecksen* (2005), United States District Court for the Northern District of Illinois Eastern Division, Civil Action No. 02 C 2180: https://www.sec.gov/litigation/litreleases/lr19351.htm

41 Harry Howard, 'Waste giant Biffa is fined £350,000 for shipping rubbish including sanitary towels, nappies and condoms to China falsely labelled as paper for recycling', *Daily Mail*, 27/09/2019: https://www.dailymail.co.uk/news/article-7512931/Biffa-fined-350-000-shipping-rubbish-including-sanitary-towels-nappies-condoms-China.html; Sandra Laville, 'UK waste firm exported "offensive" materials including used nappies', *The Guardian*, 25/06/2019: https://www.theguardian.com/environment/2019/jun/25/uk-waste-firm-exported-offensive-materials-including-used-nappies

42 'Biffa caught exporting banned waste again', UK Environment Agency, 26/07/21: https://www.gov.uk/government/news/biffa-caught-exporting-banned-waste-again

43 Amelia Gentleman, 'Three victims of trafficking and modern slavery to sue Biffa', *The Guardian*, 14/01/2021: https://www.theguardian.com/law/2021/jan/14/three-victims-of-trafficking-and-modern-slavery-to-sue-biffa

44 Goldstein, *Remains of the Everyday*, p. 408.

45 Dan Cancian, 'Malaysia Has Started Returning Tons of Trash to the West: "We Will Not Be the Dumping Ground of the World"', *Newsweek*, 28/05/2019: https://www.newsweek.com/plastic-waste-malaysia-minister-yeo-bee-bin-south-east-asia-trash-1436969

46 Vietnam and Thailand both announced bans that, at the time of writing, are due to go into effect in 2025. India has since relaxed its bans, allowing the import of some plastics, such as PET.

47 *Basel Convention on the Control of Transboundary Movements of Hazardous Wastes and their Disposal: Texts and Annexes*, Basel Convention, 2019: http://www.basel.int/TheConvention/Overview/TextoftheConvention/tabid/1275/Default.aspx

48 Global Export Data, Basel Action Network (2022): https://www.ban.org/plastic-waste-project-hub/trade-data/global-export-data-2021-annual-summary

49 For a much more eloquent discussion of this idea, including on the limitations of Mary Douglas, I recommend Max Liboiron's *Pollution Is Colonialism* (Durham, NC: Duke University Press), 2021.

50 Prior to 2021, PRNs and PERNs were known as Packaging Recovery Notes and Packaging Export Recovery Notes, respectively.

51 Kit Chellel and Wojciech Moskwa, 'A Plastic Bag's 2,000 Mile Journey Shows The Messy Truth About Recycling', *Bloomberg Businessweek*, 29/03/2022: https://www.bloomberg.com/graphics/2022-tesco-recycle-plastic-waste-pledge-falls-short/#xj4y7vzkg

52 Adam Popescu, 'The Kings of Recycling Are Fighting Over Scraps', *OneZero*, 17/09/2019: https://onezero.medium.com/the-kings-of-recycling-are-fighting-over-scraps-6cfc0a586901

53 Leslie Hook and John Reed, 'Why the world's recycling system stopped working', *Financial Times Magazine*, 25/10/2018: https://www.ft.com/content/360e2524-d71a-11e8-a854-33d6f82e62f8

54 World Bank Group, *Market Study for Malaysia: Plastics Circularity Opportunities and Barriers* (Washington, DC: World Bank), 2021: https://openknowledge.worldbank.org/handle/10986/35296

55 Karen McVeigh, 'Children as young as nine say they are ill from work recycling plastic in Turkey', *The Guardian*, 21/09/22: https://www.theguardian.com/global-development/2022/sep/21/children-as-young-as-nine-say-they-are-ill-from-work-recycling-plastic-in-turkey

4. UP IN SMOKE

1 Martin V. Melosi, *Garbage in the Cities* (Pittsburgh: University of Pittsburgh Press), p. 39.

2 Emily Cockayne, *Rummage* (London: Profile Books), 2020, p. 117.

3 Melosi, *Garbage in the Cities*, p. 158.

4 Strasser, *Waste and Want* (New York: Henry Holt and Company), 1999, p. 133.

5 They were popular in other countries too – Australia, for example.

6 Michael Holland, 'City archives show how LA banned incinerators to fight smog', KPCC.org, 13/03/2014: https://archive.kpcc.org/programs/offramp/2014/03/13/36466/city-archives-show-how-la-banned-incinerators-to-f/

7 According to DEFRA statistics. Via 'Facts & Figures', The United Kingdom Without Incineration Network, 2022 (accessed: 28/10/2022): https://ukwin.org.uk/facts

8 Beth Gardiner, 'In Europe, a Backlash is Growing Over Incinerating Garbage', *Yale Environment 360*, 01/04/2021: https://e360.yale.edu/features/in-europe-a-backlash-is-growing-over-incinerating-garbage

9   Lauren Altria, 'The Burning Problem of Japan's Waste Disposal', *Tokyo Review*, 09/07/2019: https://www.tokyoreview.net/2019/07/burning-problem-japan-waste-recycling

10 Eline Schaart, 'Denmark's "devilish" waste dilemma', *Politico*, 17/09/2020: https://www.politico.eu/article/denmark-devilish-waste-trash-energy-incineration-recycling-dilemma

11 Yun Li, Xingang Zhao, Yanbin Li and Xiaoyu Li, 'Waste incineration industry and development policies in China', *Waste Management* (2015): DOI: 10.1016/j.wasman.2015.08.008

12 Yangqing Wang, Wei Tang et al., 'Occurrence and prevalence of antibiotic resistance in landfill leachate', *Environmental Science and Pollution Research International* (2015): DOI: 10.1007/s11356-015-4514-7

13 Adam Withnall, 'Thousands of unborn foetuses incinerated to heat UK hospitals', *The Independent*, 24/03/2014: https://www.independent.co.uk/life-style/health-and-families/health-news/thousands-of-unborn-foetuses-incinerated-to-heat-uk-hospitals-9212863.html

14 At the time of writing, the clinical case into the company and its owner is still ongoing. Alex Matthews-King, 'Human body parts and organs among tonnes of medical waste stockpiled by NHS contractor', *The Independent*, 04/10/2018: https://www.independent.co.uk/news/health/nhs-human-amputated-body-parts-organs-medical-waste-hazardous-healthcare-environment-services-hsj-a8569146.html

15 Heather Rogers, *Gone Tomorrow* (New York: The New Press), 2005, p. 5.

16 'Energy for the Circular Economy: an overview of Energy from Waste in the UK', Environmental Services Association, 2018: http://www.esauk.org/application/files/7715/3589/6450/20180606_Energy_for_the_circular_economy_an_overview_of_EfW_in_the_UK.pdf

17 Josephine Moulds, 'Dirty white elephants', *SourceMaterial*, 04/02/2021: https://www.source-material.org/blog/dirty-white-elephants

18 New York State Department of Environmental Conservation, 'Matter of the Application of Covanta Energy Corporation for Inclusion of Energy from Waste Facilities as an Eligible Technology in the Main Tier of the Renewable Portfolio Standard Program. Case No. 03-E-0188', 19/08/2011: http://documents.dps.ny.gov/public/Common/ViewDoc.aspx?DocRefId={DEEA097E-A9A6-4E53-898C-0BC2F4C60CC4}

19 Silvia Candela, Andrea Ranzi et al., 'Air Pollution from Incinerators and Reproductive Outcomes', *Epidemiology* (November 2013): DOI: 10.1097/EDE.0b013e3182a712f1

20 Peiwei Xu, Yuan Chen et al., 'A follow-up study on the characterization and health risk assessment of heavy metals in ambient air particles emitted from a municipal waste incinerator in Zhejiang, China', *Chemosphere* (May 2020): DOI: 10.1016/j.chemosphere.2019.125777

21 Air Quality Consultants, 'Health Effects due to Emissions from Energy

from Waste Plant in London', Greater London Authority, May 2020: https://www.london.gov.uk/sites/default/files/gla_efw_study_final_may2020.pdf

22 Brandon Parkes, Anna Hansell, Rebecca Ghosh et al. 'Risk of congenital anomalies near municipal waste incinerators in England and Scotland: Retrospective population-based cohort study', *Environment International*, January 2020: DOI: 10.1016/j.envint.2019.05.039

23 Lijuan Zhao, Fu-Shen Zhang et al., 'Typical pollutants in bottom ashes from a typical medical waste incinerator', *Journal of Hazardous Materials* (2010): DOI: 10.1016/j.jhazmat.2009.08.066

24 Confederation of European Waste-to-Energy Plants, 'Bottom Ash Fact Sheet', 2017: https://www.cewep.eu/wp-content/uploads/2017/09/FINAL-Bottom-Ash-factsheet.pdf

25 Jack Anderson and Joseph Spear, 'The Khian Sea's Curious Voyage', *Washington Post*, 18/08/1988: https://www.washingtonpost.com/archive/business/1988/08/18/the-khian-seas-curious-voyage/da04eef4-24cc-4350-8e2e-559ba24d1be5/

26 Hope Reeves, 'A Trail of Refuse', *New York Times*, 18/02/2001: https://www.nytimes.com/2001/02/18/magazine/the-way-we-live-now-2-18-01-map-a-trail-of-refuse.html

27 In 2000, the dumped ash was finally returned from Haiti to the US for treatment and disposal. The ship's owners were prosecuted, and the ship itself eventually scrapped.

28 David Crouch, 'Nordics tackle "Achilles heel" of incineration power schemes', *Financial Times*, 11/03/2019: https://www.ft.com/content/21777666-248f-11e9-b2od-5376ca5216eb

29 'German mine to become new site for Swedish fly ash', *Bioenergy International*, 29/01/2020: https://bioenergyinternational.com/german-mine-to-become-new-site-for-swedish-fly-ash

30 Covanta, 'Ash from Energy-from-Waste' (2017): https://s3.amazonaws.com/covanta-2017/wp-content/uploads/2017/10/WP5-Ash-v1.pdf

31 Tolvik Consulting, 'UK Energy from Waste Statistics – 2020', 2021: https://www.tolvik.com/wp-content/uploads/2021/05/Tolvik-UK-EfW-Statistics-2020-Report_Published-May-2021.pdf

32 US Energy Information Administration (2021): https://www.eia.gov/tools/faqs/faq.php?id=427&t=3

33 Tolvik Consulting, 'UK Energy from Waste Statistics – 2020'.

34 Josephine Moulds, 'Dirty white elephants', *SourceMaterial*, 04/02/2021: https://www.source-material.org/blog/dirty-white-elephants

35 Will Peischel, 'Is it Time to Stop Burning Our Garbage?', *Bloomberg*, 23/05/2022: https://www.bloomberg.com/news/articles/2022-05-23/environmental-concerns-grow-over-incinerators-in-u-s

36 Global Alliance for Incinerator Alternatives, *Burning Public Money for Dirty Energy*, 2011.

37 House of Commons Treasury Committee, 'Private Finance Initiative: Written Evidence', UK Parliament, 17/05/2011: https://www.parliament.uk/globalassets/documents/commons-committees/treasury/PFI-Evidence.pdf

38 'MPs criticise "lax" schemes for waste incinerators', BBC News, 17/09/2014: https://www.bbc.co.uk/news/uk-politics-29226234

39 Altria, 'The Burning Problem of Japan's Waste Disposal'.

40 Elizabeth Royte, 'Is burning plastic waste a good idea?', *National Geographic*, 12/03/2019: https://www.nationalgeographic.com/environment/article/should-we-burn-plastic-waste

41 Nora Buli, 'Oslo commits to carbon capture plan after Fortum's $1.1 billion exit', Reuters, 22/03/2022: https://www.reuters.com/business/finlands-fortum-11-bln-deal-sell-oslo-heating-business-2022-03-22/

5. USED

1 'Key Statistics', Charity Retail Association, 2022 (accessed: 29/06/2022): https://www.charityretail.org.uk/key-statistics/

2 'Store Openings and Closures – 2021', PWC, 2021 (accessed: 30/06/22): https://www.pwc.co.uk/industries/retail-consumer/insights/store-openings-and-closures.html

3 Karine Taylor, Robin Gonzalez and Rebecca Larkin, 'Measuring the Impact of the Charitable Reuse and Recycling Sector: A Comparative study using clothing donated to charitable enterprises', MRA Consulting Group/Charitable Recycling Australia (2021): https://www.charitablerecycling.org.au/wp-content/uploads/2021/06/Charitable-Recycling-Australia-Recycled-Clothing-Impact-Assessment-240521.pdf

4 'Key Statistics', Charity Retail Association.

5 As ever in the waste business, the exact figures here are unclear.

6 Steve Eminton, 'TRI acquires Soex UK textiles recycling business', *Let's Recycle*, 09/05/2022: https://www.letsrecycle.com/news/tri-acquires-soex-uk-textiles-recycling-business/

7 Ellen MacArthur Foundation, *A New Textiles Economy: Redesigning fashion's future* (2017): http://www.ellenmacarthurfoundation.org/publications

8 Ezra Marcus, 'How Malaysia Got in on the Secondhand Clothing Boom', *New York Times*, 03/02/2022: https://www.nytimes.com/2022/02/03/style/malaysia-secondhand-clothing-grailed-etsy-ebay.html

9 Recycled fabric, historically considered of lower quality – hence the pejorative 'shoddy', meaning 'badly made'.

10 UN Comtrade data 2018; United Nations Statistics Division (2022): data. un.org (accessed: 01/07/2022).

11 I.e. $4 billion. WRAP, 'Textile Market Situation Report 2019', 2019.

12 Alden Wicker, 'Fashion has a misinformation problem. That's bad for the environment', *Vox*, 31/1/2020: https://www.vox.com/the-goods/ 2020/1/27/21080107/fashion-environment-facts-statistics-impact

13 The World Bank, 'How Much Do Our Wardrobes Cost to the Environment?', 23/09/2019: https://www.worldbank.org/en/news/ feature/2019/09/23/costo-moda-medio-ambiente

14 TRAID, 'The Impacts of Clothing: Fact Sheets' (2018): https:// traid.org.uk/wp-content/uploads/2018/09/impacts_of_clothing_ factsheet_23percent.pdf

15 Adam Minter, *Secondhand: Travels in our New Global Garage Sale* (New York: Bloomsbury), 2019, p. 15.

16 Vauhini Vara, 'Fast, Cheap and Out Of Control: Inside Shein's Sudden Rise', *WIRED*, 04/05/2022: https://www.wired.com/story/ fast-cheap-out-of-control-inside-rise-of-shein

17 Ron Gonen, *The Waste-Free World* (New York: Portfolio Penguin), 2021, p. 133.

18 Sometimes referred to as overstock, stock-in-trade, and other euphemisms.

19 Rachel Cernansky, 'Why destroying products is still an "Everest of a problem" for fashion', *Vogue Business*, 18/10/2021: https://www. voguebusiness.com/sustainability/why-destroying-products-is-still- an-everest-of-a-problem-for-fashion

20 'Three-month report: First quarter 2018', H&M Hennes & Mauritz AB (2018): https://about.hm.com/content/dam/hmgroup/groupsite/ documents/en/cision/2018/03/2145888_en.pdf. I first saw this in a *New York Times* story by Elizabeth Paton from 2018.

21 Jesper Starn, 'A Power Plant Is Burning H&M Clothes Instead of Coal', *Bloomberg*, 27/11/2017: https://www.bloomberg.com/news/articles/2017- 11-24/burning-h-m-rags-is-new-black-as-swedish-plant-ditches-coal#xj4y 7vzkg

22 Emily Chan, 'Are Your Online Returns Contributing To Fashion's Waste Problem?', *Vogue*, 07/06/2022: https://www.vogue.co.uk/fashion/article/ online-returns-landfill

23 J. P. Juanga-Labayen, I. V. Labayen and Q. Yuan, 'A Review on Textile Recycling Practices and Challenges', *Textiles,* 2022, 2, 174–188: DOI: 10.3390/textiles2010010

24 Jo Lorenz, 'Decolonising Fashion: How An Influx Of "Dead White Man's Clothes" Is Affecting Ghana', *Eco Age*, 04/08/2020: https://eco-age.com/ resources/decolonising-fashion-dead-white-mans-clothes-ghana/

25 Heather Snowden, 'This is what actually happens to your donated clothes', *High Snobiety*, 2021: https://www.highsnobiety.com/p/ kantamanto-market-the-or-circular-economy-fashion/

26   Sally Baden and Catherine Barber, 'The impact of the second-hand
     clothing trade on developing countries', Oxfam, 2005: https://
     oxfamilibrary.openrepository.com/bitstream/handle/10546/112464/
     rr-impact-second-hand-clothing-trade-developing-countries-
     010905-en.pdf

27   More commonly they are called *Mitumba*, referring to the bales they
     arrive in. Andrew Brooks, 'The hidden trade in our second-hand clothes
     given to charity', *The Guardian*, 13/02/2015: https://www.theguardian.
     com/sustainable-business/sustainable-fashion-blog/2015/feb/13/
     second-hand-clothes-charity-donations-africa

28   For a more complete history of this trade, see Andrew Brooks's *Clothing
     Poverty* (London: Bloomsbury), 2015.

29   Tragedy breeds conspiracy, but there's no denying that the market is built
     on prime real estate. 'Kantamanto fire outbreak: We're not surprised
     our market was set ablaze – traders claim', *GhanaWeb*, 16/12/2020:
     https://www.ghanaweb.com/GhanaHomePage/NewsArchive/
     Kantamanto-fire-outbreak-We-re-not-surprised-our-market-was-set-
     ablaze-Traders-claim-1135100

30   Madeleine Cobbing, Sodfa Daaji et al., 'Poisoned Gifts', Greenpeace (2022):
     https://www.greenpeace.org/static/planet4-international-stateless/
     2022/04/9f50d3de-greenpeace-germany-poisoned-fast-fashion-briefing-
     factsheet-april-2022.pdf

31   Ellen MacArthur Foundation, *A New Textiles Economy*.

32   Lucianne Tonti, 'How green are your leggings? Recycled polyester is
     not a silver bullet (yet)', *The Guardian*, 21/03/2021: https://www.
     theguardian.com/fashion/2021/mar/22/how-green-are-your-leggings-
     recycled-polyester-is not-a-silver-bullet-yet

33   'thredUp 2022 Resale Report', thredUp (2022): https://www.thredup.
     com/resale/static/2022-resaleReport-full-92a77020598ceca50f432273
     26100cc2.pdf

34   'Chile's desert dumping ground for fast fashion leftovers', *Al Jazeera*,
     08/11/2021: https://www.aljazeera.com/gallery/2021/11/8/chiles-
     desert-dumping-ground-for-fast-fashion-leftovers

6. THE CURE FOR CHOLERA

1   Deuteronomy, 23.12, *The Holy Bible: English Standard Version, Anglicised
    Edition* (London: Collins), 2007, p. 165.

2   Rose George, *The Big Necessity: Adventures in the World of Human Waste*
    (London: Portobello), 2008, p. 26.

3   Ibid.

4   Ibid., p. 11.

5   Quoted in Peter Ackroyd, *London: The Biography* (London: Vintage), 2001, p. 339.

6   Stephen Halliday, *An Underground Guide to Sewers or: Down, Through & Out in Paris, London, New York &c.* (London: Thames & Hudson), 2019, p. 49.

7   Ibid.

8   George, *The Big Necessity*, p. 27.

9   Halliday, *An Underground Guide to Sewers*, p. 36.

10 Ibid., p. 53.

11 Steven Johnson, *The Ghost Map: A Street, A City, an Epidemic, and the Hidden Power of Urban Networks* (London: Penguin Books), 2008, p. 12.

12 John Bugg, 'On the origin of faeces', *Times Literary Supplement*, 27/10/2017: https://www.the-tls.co.uk/articles/on-the-origin-of-faeces-stink/

13 George, *The Big Necessity*, p. 2.

14 Stephen Halliday and Adam Hart-Davis, *The Great Stink of London: Sir Joseph Bazalgette and the Cleaning of the Victorian Metropolis* (Stroud: The History Press), p. 263 (eBook edition).

15 Ackroyd, *London: The Biography*, p. 344.

16 Benjamin Disraeli, *Hansard*, HC Deb 15 July 1858, Vol. 151: https://api.parliament.uk/historic-hansard/commons/1858/jul/15/first-reading

17 Today, Beckton and Crossness are parts of Outer London, in the boroughs of Newham and Bexley, respectively.

18 Halliday, *An Underground Guide to Sewers*, p. 110.

19 Institute of Civil Engineers, 'London sewer system', 2022: https://www.ice.org.uk/what-is-civil-engineering/what-do-civil-engineers-do/london-sewer-system

20 Ackroyd, *London: The Biography*, p. 344.

21 I first heard this story on the *99% Invisible* podcast, specifically in the 2013 episode 'Reversal of Fortune'. See also Halliday, *An Underground Guide to Sewers*.

22 E. Ashworth Underwood, 'The History of Cholera in Great Britain', *Proceedings of the Royal Society of Medicine*, Vol. XLI (1947), p. 170.

23 George, *The Big Necessity*, p. 3.

24 World Health Organization, 'Sanitation: Key Facts', 2022 (accessed: 26/09/22): https://www.who.int/news-room/fact-sheets/detail/sanitation

25 *Diarrhea: Common Illness, Global Killer*, Centers For Disease Control And Prevention, p. 1.

26 Jamie Bartram and Sandy Cairncross, 'Hygiene, Sanitation, and Water: Forgotten Foundations of Health', *PLoS Med* (2010): DOI: 10.1371/journal.pmed.1000367

27 'London's Population', Greater London Authority, 2021 (accessed: 26/09/22): https://data.london.gov.uk/dataset/londons-population

28 'The Sewage Treatment Process', Thames Water, 2022: https://www.

thameswater.co.uk/about-us/responsibility/education/the-sewage-treatment-process

29   Jamie Grierson, 'Wet wipes "forming islands" across UK after being flushed', *The Guardian*, 02/11/2021: https://www.theguardian.com/environment/2021/nov/02/wet-wipes-forming-islands-across-uk-after-being-flushed

30   Lamiat Sabin, 'Wet wipe island the size of two tennis courts has "changed flow of River Thames"', *The Independent*, 24/06/2022: https://www.independent.co.uk/climate-change/news/wet-wipes-river-thames-waste-hammersmith-b2108791.html

31   Sandra Laville, 'Raw sewage discharged into English rivers 375,000 times by water firms', *The Guardian*, 31/03/2022: https://www.theguardian.com/environment/2022/mar/31/sewage-released-into-english-rivers-for-27m-hours-last-year-by-water-firms

32   Stephanie L. Wear et al., 'Sewage pollution, declining ecosystem health, and cross-sector collaboration', *Biological Conservation*, Vol. 255 (2021): DOI: 10.1016/j.biocon.2021.109010

33   Annabel Ferriman, 'BMJ readers choose the "sanitary revolution" as the greatest medical advance since 1840', *British Medical Journal* (2007): DOI: 10.1136/bmj.39097.611806.DB

34   United Nations, 'The United Nations World Water Development Report 2018', 2018: https://www.unwater.org/publications/world-water-development-report-2018

7. A THIRD OF EVERYTHING

1   United Nations Environment Programme, 'Food Waste Index Report 2021', 2021: https://www.unep.org/resources/report/unep-food-waste-index-report-2021

2   As with anything involving waste, these figures are extremely difficult to calculate, and vary widely. The UN estimates only 17 per cent of food is wasted, but does not include farm wastage; a recent study by the World Wildlife Fund and Tesco, a British retailer, estimated the number may be as high as 40 per cent, or 2.5 billion tonnes: *Driven to Waste: The Global Impact of Food Loss and Waste on Farms*, World Wildlife Fund, August 2021: https://wwfint.awsassets.panda.org/downloads/wwf_uk__driven_to_waste____the_global_impact_of_food_loss_and_waste_on_farms.pdf

3   Intergovernmental Panel on Climate Change, 'Climate Change and Land: an IPCC special report on climate change, desertification, land degradation, sustainable land management, food security, and greenhouse gas fluxes in terrestrial ecosystems', 2019: https://www.ipcc.ch/srccl/chapter/chapter-5/

4   UN Food and Agriculture Organization, 'Food wastage footprint: Impacts on natural resources', 2013: https://www.fao.org/3/i3347e/i3347e.pdf

5   'China wastes almost 30% of its food', *Nature*, 15/07/2021: https://www.nature.com/articles/d41586-021-01963-3

6   UNEP, *Food Waste Index Report 2021*, p. 36.

7   US Environmental Protection Agency, 'Food: Material-Specific Data' (2018): https://www.epa.gov/facts-and-figures-about-materials-waste-and-recycling/food-material-specific-data

8   An exception being largely Muslim Xinjiang, which is known for its nose-to-tail meat eating, as Tristram Stuart writes in *Waste: The Global Food Scandal*.

9   Helen Davidson, 'China to bring in law against food waste with fines for promoting overeating', *The Guardian*, 23/12/2020 https://www.theguardian.com/world/2020/dec/23/china-to-bring-in-law-against-food-waste-with-fines-for-promoting-overeating

10 WRAP, 'Food surplus and waste in the UK – key facts', October 2021: https://wrap.org.uk/sites/default/files/2021-10/food-%20surplus-and-%20waste-in-the-%20uk-key-facts-oct-21.pdf

11 Kylie Ackers, 'How Much Bread Do We Waste in the UK?', *Eco & Beyond* (2019): https://www.ecoandbeyond.co/articles/much-bread-waste-uk

12 Amy Halloran, 'How Do We Solve America's Oversupply of Bread?', *Eater*, 01/09/2017: https://www.eater.com/2017/9/1/16239964/bread-excess-waste-production-problem-solution

13 Rebecca Smithers, 'UK households waste 4.5m tonnes of food each year', *The Guardian*, 24/01/2020.

14 Although some charities believe this number to be higher. Brigid Francis-Devine, Shadi Danechi, Yago Zayed et al., 'Food Poverty: Households, food banks and free school meals' (London: UK House Of Commons Library), 2022: https://researchbriefings.files.parliament.uk/documents/CBP-9209/CBP-9209.pdf

15 Or just not cutting forests down in the first place. The UN's Food and Agriculture Organization calculates that 80 per cent of deforestation is for agriculture, which in turn is used to produce food that we never eat. United Nations Food and Agriculture Organization, 'Food wastage footprint: impacts on natural resources', 2013.

16 *Driven to Waste*, World Wildlife Fund.

17 Tristram Stuart, *Waste: Uncovering the Global Food Scandal* (London: Penguin), 2009, p. 89.

18 'BOGOF deals exacerbate food waste problem', *Let's Recycle*, 13/01/2014: https://www.letsrecycle.com/news/bogof-deals-exacerbate-food-waste-problem/

19 Chandrima Shrivastava et al., 'To wrap or to not wrap cucumbers?' (pre-print), *Engrxiv*, 30/07/2021: DOI: 10.31224/osf.io/dyx9b

20 WRAP, 'Reducing household food waste and plastic packaging:
    Evidence and Insights' (2022): https://wrap.org.uk/resources/report/
    reducing-household-food-waste-and-plastic-packaging
21 Ashifa Kassam, 'Spain fights food waste with supermarket fines and doggy
    bags', *The Guardian*, 07/06/2022: https://www.theguardian.com/
    world/2022/jun/07/spain-fights-food-waste-with-supermarket-fines-and-
    doggy-bags
22 Rosemary H. Jenkins et al., 'The relationship between austerity and food
    insecurity in the UK: A systematic review', *EClinicalMedicine* 33 (2021),
    DOI: 10.1016/j.eclinm.2021.100781
23 Feeding America, 'Annual Report 2021', 2021: FA_2021AnnReport_
    FULL_d7_final.pdf (feedingamerica.org)
24 Another helpful euphemism, denoting food on, or past its sell-by date. (In
    technical jargon, this food isn't 'wasted' because it's being used as animal
    feed, energy, or fertiliser.)
25 FareShare, 'The Wasted Opportunity', 2018: https://fareshare.org.uk/
    wp-content/uploads/2018/10/J3503-Fareshare-Report_aw_no_crops.pdf
26 Gregory A. Baker, Leslie C. Gray et al., 'On-farm food loss in northern and
    central California: Results of field survey measurements', *Resources, Conservation
    and Recycling* (October 2019): DOI: 10.1016/j.resconrec.2019.03.022
27 Anna Sophie Gross, 'One in six pints of milk thrown away each year,
    study shows', *The Guardian*, 28/11/2018: https://www.theguardian.com/
    environment/2018/nov/28/one-in-six-pints-of-milk-thrown-away-each-
    year-study-shows
28 UN Food and Agriculture Organization, 'The State of World Fisheries and
    Aquaculture 2020. Sustainability in action', 2020: DOI: 10.4060/
    ca9229en
29 *Driven to Waste*, World Wildlife Fund.
30 Feedback, 'Farmers Talk Food Waste: supermarkets' role in crop
    waste on UK farms' (2018): https://feedbackglobal.org/wp-content/
    uploads/2018/08/Farm_waste_report_.pdf
31 Melissa Chan, 'Dairy Farmers Pour Out 43 Million Gallons Of Milk
    Due To Surplus', TIME, 13/10/2016: https://time.com/4530659/
    farmers-dump-milk-glut-surplus/
32 Frank Trentmann, *Empire of Things: How We Became a World of Consumers, from
    the Fifteenth Century to the Twenty-First* (London: Penguin), 2016, p. 649.
33 Fellow food-nerds can read them here: The European Union,
    'Commission Implementing Regulation (EU) No 543/2011 of 7 June
    2011', 15/11/2021: https://eur-lex.europa.eu/legal-content/EN/
    TXT/?uri=CELEX:32011R0543
34 Stephen Castle, 'EU relents and lets a banana be a banana', *New York Times*,
    12/11/2008: https://www.nytimes.com/2008/11/12/world/europe/12iht-
    food.4.17771299.html

35 It's partially true. The EU once did have a detailed commercial grading standard for bananas, which dictated that some shapes were only to be sold for food manufacturing, not retail. It was abandoned in 2008, although the myth, which returned during the 2016 Brexit vote, persists. The EU still does issue marketing standards for several foods, including tomatoes, apples, pears, strawberries, and citrus fruit.

36 Feedback, 'Farmers Talk Food Waste'.

37 T. J. McKenzie, L. Singh-Peterson and S. J. R. Underhill, 'Quantifying Postharvest Loss and the Implication of Market-Based Decisions: A Case Study of Two Commercial Domestic Tomato Supply Chains in Queensland, Australia', *Horticulturae* (2017): DOI: 10.3390/horticulturae3030044

38 GCA Annual Survey 2020: https://assets.publishing.service.gov.uk/government/uploads/system/uploads/attachment_data/file/886761/GCA_YouGov_2020_Presentation.pdf

39 Rebecca Smithers, 'Pumpkin waste in UK predicted to hit scary heights this Halloween', *The Guardian*, 23/10/2019: https://www.theguardian.com/environment/2019/oct/23/pumpkin-waste-uk-halloween-lanterns

40 When measured by the share of food served that goes to waste, hospitality and food service are actually the most wasteful sector. But what we eat out is dwarfed by what we cook and eat at home.

41 During the reporting of this book, many UK supermarkets have begun to abandon Best Before dates, although internationally progress on this issue varies.

42 William Rathje and Cullen Murphy, *Rubbish! The Archaeology of Garbage*, (New York: HarperCollins), 1992, p. 62.

43 Full credit to Tristram Stuart for that ingenious method.

8. BREAKDOWN

1 Likely something of an exaggeration.

2 S. Kaza, L. Yao, P. Bhada-Tata and F. Van Woerden, 'What A Waste 2.0: A Global Snapshot of Solid Waste Management to 2050', The World Bank, 2018, p. 34: DOI: 10.1596/978-1-4648-1329-0

3 Figures according to WRAP. See 'Anaerobic Digestion and Composting Industry Market Survey Report 2020', WRAP (2020): https://wrap.org.uk/resources/report/anaerobic-digestion-and-composting-latest-industry-survey-report-new-summaries

4 US Environmental Protection Agency, 'Preventing Wasted Food At Home', 2022: https://www.epa.gov/recycle/preventing-wasted-food-home

5 Steven Johnson, *The Ghost Map: A Street, A City, An Epidemic and the Hidden Power of Urban Networks* (London: Penguin), 2008, p. 8.

6 Many people might use the cliché 'highly efficient' here, but as we've

seen in our discussion of food waste, that simply isn't true of modern agriculture.

7   Karl Marx, *Capital, Vol. I* (New York: Vintage), 1967, p. 637.

8   George Monbiot, 'The secret world beneath our feet is mind-blowing – and the key to our planet's future', *The Guardian*, 07/05/2022: https://www. theguardian.com/environment/2022/may/07/secret-world-beneath-our-feet-mind-blowing-key-to-planets-future

9   George Monbiot, *Regenesis: Feeding the World Without Devouring the Planet* (London: Allen Lane), 2022, p. 23.

10 Ibid., p. 72.

11 National Oceanic and Atmospheric Administration, 'Larger-than-average Gulf of Mexico "Dead Zone" measured', 03/08/2021: https://www.noaa. gov/news-release/larger-than-average-gulf-of-mexico-dead-zone-measured

12 Jonathan Watts, 'Third of Earth's soil is acutely degraded due to agriculture', *The Guardian*, 12/09/2017: https://www.theguardian.com/environment/2017/ sep/12/third-of-earths-soil-acutely-degraded-due-to-agriculture-study

13 Aaron Sidder, 'The Green, Brown, and Beautiful Story of Compost', *National Geographic*, 09/09/2016: https://www.nationalgeographic.com/ culture/article/compost--a-history-in-green-and-brown

14 Heather Rogers, *Gone Tomorrow: The Hidden Life of Garbage* (New York: The New Press), 2005, p. 85.

15 Moises Velasquez-Manoff, 'Can Dirt Save The Earth?', *New York Times*, 18/04/2018: https://www.nytimes.com/2018/04/18/magazine/dirt-save-earth-carbon-farming-climate-change.html

16 Nicole E. Tautges, Jessica L. Chiartas et al., 'Deep soil inventories reveal that impacts of cover crops and compost on soil carbon sequestration differ in surface and subsurface soils', *Global Change Biology* 25(11), 2019: DOI: 10.1111/gcb.14762

17 The viability of biochar as a carbon sequestration method is disputed.

18 Damian Carrington, 'Third of all compost sold in UK is climate-damaging peat', *The Guardian*, 31/03/2022: https://www.theguardian.com/ environment/2022/mar/31/third-of-all-compost-sold-in-uk-is-climate-damaging-peat

19 'Waste Strategy: Implications for local authorities: Government Response to the Committee's Nineteenth Report of Session 2017–19', UK Parliament, 19/05/2020: https://publications.parliament.uk/pa/cm5801/cmselect/ cmcomloc/363/36302.htm

20 'Suntory Introduces 100% Plant-Based PET Bottle Prototypes', Suntory. com, 12/03/2021: https://www.suntory.com/news/article/14037E.html

21 Simon Hann, Sarah Ettlinger et al., 'The Impact of the Use of "Oxo-degradable" Plastic on the Environment', EUNOMIA for the European Commission, 07/08/2016: https://op.europa.eu/en/publication-detail/-/ publication/bb3ec82e-9a9f-11e6-9bca-01aa75ed71a1

22 Darrel Moore, 'UK Government may ban oxo-degradable plastics following consultation', *Circular*, 12/04/2021: https://www.circularonline. co.uk/news/uk-government-may-ban-oxo-degradable-plastics-following-consultation/

23 *Standards and Regulations for the Bio-based Industry STAR4BBI*, European Commission, 2019.

24 Susan Freinkel, *Plastic: A Toxic Love Story* (New York: Houghton Mifflin Harcourt), 2011, p. 214.

25 Nick Hughes and David Burrows, 'Footprint Investigation: Parliament burnt by compostable pledge', *Footprint*, 07/07/2019: https://www. foodservicefootprint.com/footprint-investigation-parliament-burnt-by-compostable-pledge/

26 Imogen Napper and Richard C. Thompson, 'Environmental Deterioration of Biodegradable, Oxo-biodegradable, Compostable, and Conventional Plastic Carrier Bags in the Sea, Soil, and Open-Air Over a 3-Year Period', *Environmental Science & Technology* (2019): DOI: 10.1021/acs.est.8b06984

27 UK Government, 'Official Statistics Section 3: Anaerobic Digestion', 9/12/2021 (accessed: 14/07/2022): https://www.gov.uk/government/statistics/area-of-crops-grown-for-bioenergy-in-england-and-the-uk-2008-2020/section-3-anaerobic-digestion

28 Tristram Stuart, *Waste: Uncovering the Global Food Scandal* (London: Penguin), 2009, p. 233.

29 I first read this in Stuart, *Waste*, p. 233. Updated figures from Maxime Lemonde, 'Sweden: A Pioneer in Natural Gas Vehicle, NGV and BIONGV', *Biogas World*, 19/06/2018: https://www.biogasworld.com/news/vehicle-ngv-biongv-sweden/

30 Stuart, *Waste*, p. 241.

31 UK Government, 'Official Statistics Section 3: Anaerobic Digestion', 09/12/2021.

32 Nils Klawitter, 'Biogas Boom in Germany Leads to Modern-Day Land Grab', *Der Spiegel*, 30/08/2012: https://www.spiegel.de/international/germany/biogas-subsidies-in-germany-lead-to-modern-day-land-grab-a-852575.html

33 Ewa Wiśniowska, Anna Grobelak et al., 'Sludge legislation comparison between different countries', *Industrial and Municipal Sludge: Emerging Concerns and Scope for Resource Recovery* (2019): DOI: 10.1016/B978-0-12-815907-1.00010-6Get

34 Julia Rosen, 'Humanity is flushing away one of life's essential elements', *The Atlantic*, 08/02/2021: https://www.theatlantic.com/science/archive/2021/02/phosphorus-pollution-fertiliser/617937/

35 Hannatou O. Moussa, Charles I. Nwankwo et al., 'Sanitized human urine (Oga) as a fertiliser auto-innovation from women farmers in Niger', *Agronomy for Sustainable Development* (2021): DOI: 10.1007/s13593-021-00675-2

36 Chelsea Wald, 'The urine revolution: how recycling pee could help to

save the world', *Nature*, 09/02/2022: https://www.nature.com/articles/
d41586-022-00338-6

37 Yara, 'Veolia and Yara partner to propel European circular economy',
    21/01/2019: https://www.yara.com/corporate-releases/veolia-and-yara-
    partner-to-propel-european-circular-economy/

38 Monbiot, *Regenesis*, p. 66.

39 Tom Perkins, 'Forever chemicals may have polluted 20m acres of US
    cropland, study says', *The Guardian*, 08/05/2022: https://www.theguardian.
    com/environment/2022/may/08/us-cropland-may-be-contaminated-
    forever-chemicals-study

40 Youn Jeong Choi, Rooney Kim Lazcano et al., 'Perfluoroalkyl Acid
    Characterization in U.S. Municipal Organic Solid Waste Composts',
    Environmental Science & Technology Letters 6(6), 2019, pp. 372–7: DOI:
    10.1021/acs.estlett.9b00280

41 Meththika Vithanage, Sammani Ramanayaka et al., 'Compost as a carrier
    for microplastics and plastic-bound toxic metals into agroecosystems',
    Current Opinion in Environmental Science & Health, Vol. 24 (2021): DOI:
    10.1016/j.coesh.2021.100297

9. UNHOLY WATER

1 In the Yamuna, the problem is reinforced by water hyacinths feeding on the
   nutrient-rich sewage, which bloom and die off in the river upstream, their
   decay releasing further phosphates and surfactant-like chemicals.

2 Suaibu O. Badmus et al., 'Environmental risks and toxicity of surfactants:
   overview of analysis, assessment, and remediation techniques', *Environmental
   Science and Pollution Research* (2021): DOI: 10.1007/s11356-021-16483-w

3 Anita Singh, Sudesh Chaudhary and Brij Dehiya, 'Metal Content in the
   Surface Water of Yamuna River, India', *Applied Ecology and Environmental
   Sciences* 8(5), January 2022, pp. 244–253,: DOI: 10.12691/aees-8-5-9

4 Paras Singh and Jasjeev Gandhiok, 'At fault for a frothy Yamuna: Raw
   sewage, frothing agents', *Hindustan Times*, 09/11/2021: https://www.
   hindustantimes.com/cities/delhi-news/at-fault-for-a-frothy-yamuna-raw-
   sewage-frothing-agents-101636396801259.html

5 Unbelievably, this foul record is actually a major improvement on previous
   years, where faecal coliform has reached nearly half a million times the
   Indian bathing water standards, according to Victor Mallet in *River of Life,
   River of Death* (New York: Oxford University Press), 2017.

6 Kushagra Dixit, 'Yamuna water quality worsening, shows UPPCB data',
   Hindustan Times, 11/07/2021: https://www.hindustantimes.com/cities/
   noida-news/yamuna-water-quality-worsening-shows-uppcb-data-
   101626027638742

7   Chetan Chauhan, 'Yamuna a dead river, says report, even as focus on Clean Ganga', *Hindustan Times*, 18/04/2019: https://www.hindustantimes.com/delhi/yamuna-a-dead-river-says-report-even-as-focus-on-clean-ganga/story-4R6VXEcjNOlLSelnREqrxN

8   Central Pollution Control Board, 'Water Quality Data of Rivers Under National Water Quality Monitoring Programme (NWMP)', Ministry of Environment, Forest and Climate Change, Government Of India, 2020: https://cpcb.nic.in/wqm/2020/WQuality_River-Data-2020.pdf

9   There are notable exceptions to this, including newborns, holy men, and some Dalits, who are buried.

10   Anuja Jaiswal, 'Thousands of dead fish wash ashore on Yamuna banks in Agra', *Times of India*, 27/07/2021: https://timesofindia.indiatimes.com/city/agra/thousands-of-dead-fish-wash-ashore-on-banks-in-agra/articleshow/84769550.cms

11   As well as India, the Ganges basin also takes in parts of Nepal and Tibet.

12   WWF India, *India's Mitras: Friends of the River*, World Wildlife Fund (2017): https://www.wwf.org.uk/sites/default/files/2017-06/170616_Ganga_Mitras_CS-external.pdf

13   There are other versions of this story, Hinduism being rich with creation myth.

14   George Black, 'What It Takes To Clean The Ganges', *New Yorker*, 25/07/2016: https://www.newyorker.com/magazine/2016/07/25/what-it-takes-to-clean-the-ganges

15   Simon Scarr, Weiyi Cai et al., 'The Race To Save The River Ganges', Reuters, 18/01/2019: https://graphics.reuters.com/INDIA-RIVER/010081TW39P/index.html

16   '78% of sewage generated in India remains untreated', *Down To Earth*, 05/04/2016: https://www.downtoearth.org.in/news/waste/-78-of-sewage-generated-in-india-remains-untreated-53444

17   Scarr et al., 'The Race To Save The River Ganges'.

18   Estimates on exactly how much vary, but pledges are close to £2.47 billion. However, as of 2019, much of this budget remained unspent.

19   Black, 'What It Takes To Clean The Ganges'.

20   Lorraine Boissoneault, 'The Cuyahoga River Caught Fire at Least a Dozen Times, but No One Cared Until 1969', *Smithsonian Magazine*, 19/06/2019: https://www.smithsonianmag.com/history/cuyahoga-river-caught-fire-least-dozen-times-no-one-cared-until-1969-180972444/

21   David Stradling and Richard Stradling, 'Perceptions of the Burning River: Deindustrialization and Cleveland's Cuyahoga River', *Environmental History*, Vol. 13 (2008): http://www.jstor.org/stable/25473265

22   Wes Siler, '51 Years Later, the Cuyahoga River Burns Again', *Outside*, 28/08/2020: https://www.outsideonline.com/outdoor-adventure/environment/cuyahoga-river-fire-2020-1969

23   Michael C. Newman and Michael A. Unger, *Fundamentals of Ecotoxicology* (Abingdon: Taylor & Francis), 2003, p. 3.

24   As noted by Max Liboiron, not everyone uses this term (the US EPA and NOAA use 'accommodative capacity', for example) but the underlying principle remains the same.

25   The UK's Department for Environment, Food & Rural Affairs refers to these as 'Maximum Allowable Concentration', the US's EPA prefers 'Total Maximum Daily Loads'. Choose your poison.

26   Water, England and Wales: The Water Supply (Water Quality) Regulations 2016. Available at: https://www.legislation.gov.uk/uksi/2016/614/pdfs/uksi_20160614_en.pdf.

27   Max Liboiron, *Pollution Is Colonialism* (Durham, NC, and London: Duke University Press), 2021, p. 5.

28   Jocelyn Kaiser, 'New Mercury Reports Supports Stringent Safety Levels', *Science*, 14/07/2000: https://www.science.org/content/article/new-mercury-report-supports-stringent-safety-levels

29   Joseph Winters, 'European regulators propose "dramatic" new regulation for BPA', *Grist*, 21/12/2021: https://grist.org/regulation/europe-proposes-dramatic-new-regulation-for-bpa/

30   IARC, 'Agents Classified by the IARC Monographs, Volumes 1–132': https://monographs.iarc.who.int/agents-classified-by-the-iarc (accessed: 13/06/2022).

31   European Chemicals Agency, 'REACH Registration Statistics, All Countries': https://echa.europa.eu/documents/10162/2741157/registration_statistics_en.pdf/58c2d7bd-2173-4cb9-eb3b-a6bc14a6754b?t=1649160655122 (accessed: 13/06/2022).

32   Dr Hélène Loonen, Dolores Romano, Tatiana Santos and Elise Vitali, *Chemical Evaluation: Achievements, challenges and recommendations after a decade of REACH*, European Environmental Bureau, 2019, p. 5

33   The UK, post-Brexit, is due to pass its own version of REACH, but has repeatedly extended the deadline for companies to register their safety data (right now it is in 2025), and has suggested the legislation may be watered down, because the process is too complicated.

34   The CAS database, created by the American Chemical Society, lists more than 264 million registered chemical substances – although this includes a register of all known biomedical chemicals, many of which are not in circulation.

35   Zhanyun Wang, Glen W. Walker et al., 'Toward a Global Understanding of Chemical Pollution: A First Comprehensive Analysis of National and Regional Chemical Inventories', *Environmental Science & Technology* (2020): DOI: 10.1021/acs.est.9b06379

36   Linn Persson et al., 'Outside the Safe Operating Space of the Planetary Boundary for Novel Entities', *Environmental Science & Technology* (2022): DOI: 10.1021/acs.est.1c04158

37   I'm not usually one for such claims, which tend to be based on belief as

much as maths (particularly when the impact of chemical pollutants such as microplastics at scale are poorly understood). But you can understand the concern.

38   Mohammad Faisal, 'After 240 years Kanpur leather industry on the verge of permanent closure: An annual loss of 12000 crores', TwoCircles.net, 28/05/2019: https://twocircles.net/2019may28/431522.html

39   Preeti Trivedi, *Kanpur Unveiled* (Kanpur: Bhartiya Sahitya Inc), 2017, pp. 38–46.

40   Abhishek Waghmare, 'The slow death of Kanpur's leather economy and UP's job crisis', *Business Standard*, 11/02/2017: https://www.business-standard.com/article/economy-policy/the-slow-death-of-kanpur-s-leather-economy-and-up-s-job-crisis-117021100176_1.html

41   '10 Indian cities top WHO list of most polluted in the world', Associated Press, 03/05/2018: https://apnews.com/article/77ba40f9061e44cbb9a9c0984d454698

42   As measured by PM2.5. When I was there, it broke 145.1μg/m³ according to the website IQ Air.

43   François Jarrige and Thomas Le Roux, *The Contamination of the Earth: A History of Pollutions in the Industrial Age* (Cambridge: MIT Press), 2020, pp. 824–38 (Kindle iOS edition).

44   Katarzyna Chojnacka, Dawid Skrzypczak et al., 'Progress in sustainable technologies of leather wastes valorization as solutions for the circular economy', *Journal of Cleaner Production*, Vol. 313 (2021): DOI: 10.1016/j.jclepro.2021.127902

45   Kazi Madina Maraz, 'Benefits and problems of chrome tanning in leather processing: Approach a greener technology in leather industry', *Materials Engineering Research* 3(1), 2001, pp. 156–64: DOI: 10.25082/MER.2021.01.004

46   Yujiao Deng et al., 'The Effect of Hexavalent Chromium on the Incidence and Mortality of Human Cancers: A Meta-Analysis Based on Published Epidemiological Cohort Studies', *Frontiers in Oncology*, 04/02/2019: DOI: 10.3389/fonc.2019.00024

47   Ajoy Ashirwad Mahaprashasta, 'As Kanpur Tanneries Face Extinction, Adityanath's (Mis)Rule Dominates Poll Talk', *The Wire*, 19/02/2022: https://thewire.in/labour/as-kanpur-tanneries-face-extinction-adityanaths-misrule-dominates-poll-talk

48   Malavika Vyawahare, 'This Kanpur village drinks neon green water & lives near a toxic waste dump as big as CP', *The Print*, 15/03/2019: https://theprint.in/india/this-kanpur-village-drinks-neon-green-water-lives-near-a-toxic-waste-dump-as-big-as-cp/205769/

49   Dipak Paul, 'Research on heavy metal pollution of river Ganga: A review', *Annals of Agrarian Science* 15(2), 2017, pp. 278–86: DOI: 10.1016/j.aasci.2017.04.001

50   Iqbal Ahmad and Sadhana Chaurasia, 'Study on Heavy Metal Pollution in
     Ganga River at Kanpur (UP)', *Journal of Emerging Technologies and Innovative
     Research*, 6 (2019), pp. 391–8.

51   Tanuj Shukla, Indra Sen et al., 'A Time-Series Record during Covid-19
     Lockdown Shows the High Resilience of Dissolved Heavy Metals in the
     Ganga River', *Environmental Science & Technology Letters* 8(4), 2021, pp. 301–6:
     DOI: 10.1021/acs.estlett.0c00982

52   The other 'Big Pollution Diseases' were Minamata disease (methylmercury
     poisoning), Niigata Minamata disease (also methylmercury poisoning)
     and Yokkaichi asthma (sulphur dioxide poisoning). Michael C. Newman
     and Michael A. Unger, *Fundamentals of Ecotoxicology – Second Edition* (Boca
     Raton: Lewis Publishers), 2002, p. 3.

53   The United States, while a signatory, has never ratified the treaty.

54   Gabriel Dunsmith, 'Silent Spring comes to life in DDT-stricken
     town', *Greenwire*, 02/02/2017: https://www.eenews.net/articles/
     silent-spring-comes-to-life-in-ddt-stricken-town/

55   Alan J. Jamieson, Tamas Malkocs et al., 'Bioaccumulation of persistent
     organic pollutants in the deepest ocean fauna', *Nature Ecology & Evolution*
     (2017): DOI: 10.1038/s41559-016-0051

56   Brigit Katz, 'U.K. Killer Whale Contained Staggering Levels of
     Toxic Chemical', *Smithsonian Magazine*, 04/05/2017: https://www.
     smithsonianmag.com/smart-news/body-uk-orca-contained-staggering-
     levels-toxic-chemical-180963135/

57   Rebecca Giggs, *Fathoms: The World in The Whale* (London: Scribe UK), 2021,
     p. 11.

58   For what I think is the most essential modern history of PFASs, see Sharon
     Lerner's 'Bad Chemistry' series at the *Intercept*. 'Bad Chemistry', *The
     Intercept* (2019): https://theintercept.com/collections/bad-chemistry/

59   CHEM Trust, *PFAS: the "Forever Chemicals"*, 2019: https://chemtrust.org/wp-
     content/uploads/PFAS_Brief_CHEMTrust_2019.pdf

60   Sharon Lerner, 'The Teflon Toxin: The Case Against DuPont', *The
     Intercept*, 17/08/2015: https://theintercept.com/2015/08/2017/
     teflon-toxin-case-against-dupont/

61   Nathaniel Rich, 'The Lawyer Who Became DuPont's Worst Nightmare',
     New York Times, 06/01/2016: https://www.nytimes.com/2016/01/10/
     magazine/the-lawyer-who-became-duponts-worst-nightmare.html

62   Sharon Lerner, 'Chemours is using the US as an unregulated dump
     for Europe's toxic GenX waste', *The Intercept*, 01/02/2019: https://
     theintercept.com/2019/02/01/chemours-genx-north-carolina-netherlands/

63   Ali Alavian-Ghavanini and Joëlle Rüegg, 'Understanding Epigenetic Effects
     of Endocrine Disrupting Chemicals: From Mechanisms to Novel Test
     Methods', *Basic & Clinical Pharmacology & Toxicology* 112(1), 2018, pp. 38–45:
     DOI: 10.1111/bcpt.12878

64 Linda G. Kahn, Claire Philippat et al., 'Endocrine-disrupting chemicals: implications for human health', *The Lancet Diabetes & Endocrinology* (2020): DOI: 10.1016/S2213-8587(20)30129-7

65 Andrea Di Nisio, Iva Sabovic et al., 'Endocrine disruption of androgenic activity by perfluoroalkyl substances: clinical and experimental evidence', *Journal of Clinical Endocrinology & Metabolism* (2019): DOI: 10.1210/jc.2018-01855

66 Andreas Kortenkamp, Martin Scholze et al., 'Combined exposures to bisphenols, polychlorinated dioxins, paracetamol, and phthalates as drivers of deteriorating semen quality', *Environment International* (2022): DOI: 10.1016/j.envint.2022.107322

67 Credit here: Max Liboiron makes this argument in more detail in their excellent *Pollution Is Colonialism* (2021).

68 Tom Perkins, 'EPA imposes stricter limits on four types of toxic "forever chemicals"', *The Guardian*, 15/06/2022: https://www.theguardian.com/environment/2022/jun/15/epa-limits-toxic-forever-chemicals

69 Ibid.

70 Waste Dive, 'Disposing "forever" toxics: How the waste and recycling industry is tackling the PFAS chemicals crisis', 2021: https://www.wastedive.com/news/pfas-chemicals-toxic-disposal-waste-crisis/587045/

10. CONTROL, DELETE

1 Vanessa Forti, Cornelis Peter Baldé et al., *The Global E-waste Monitor 2020: Quantities, flows, and the circular economy potential*, United Nations University and United Nations Institute for Training and Research, 2020.

2 Simon O'Dea, 'Number of smartphones sold to end users worldwide from 2007 to 2021 (in million units)', *Statista*, 06/05/2022 (accessed: 24/06/2022): https://www.statista.com/statistics/263437/global-smartphone-sales-to-end-users-since-2007/

3 'Global PC shipments pass 340 million in 2021 and 2022 is set to be even stronger', *Canalys*, 12/01/2022: https://www.canalys.com/newsroom/global-pc-market-Q4-2021

4 Saranraj Mathivanan, 'Global Headphones Market Shipped nearly 550 million units in 2021', Futuresource Consulting (2022): https://www.futuresource-consulting.com/insights/global-headphones-market-shipped-nearly-550-million-units-in-2021

5 Calculated using Apple, 'Environmental Responsibility Report: 2019 Progress Report, covering fiscal year 2018', 2019: https://www.apple.com/environment/pdf/Apple_Environmental_Responsibility_Report_2019.pdf

6 Bianca Nogrady, 'Your old phone is full of untapped precious metals', *BBC Future*, 18/10/2016: https://www.bbc.com/future/article/20161017-your-old-phone-is-full-of-precious-metals

7   Adam Minter, *Junkyard Planet: Travels in the Billion-Dollar Trash Trade* (New York: Bloomsbury Press), 2013, p. 175.

8   World Economic Forum, 'A New Circular Vision for Electronics: Time For A Global Reboot', 2019, p. 5.

9   Lauren Joseph and James Pennington, 'Tapping the economic value of e-waste', *China Daily*, 29/10/2018: http://europe.chinadaily.com.cn/a/201810/29/WS5bd64e5aa310eff3032850ac.html

10 World Economic Forum, 'A New Circular Vision', p. 11.

11 Or $57 billion, according to Forti, Baldé et al., *The Global E-waste Monitor 2020*.

12 Originally, the company was called Computer Recyclers of America, but it rebranded and relaunched in 2005 as ERI.

13 'Nonferrous Scrap Terminology', *Recycling Today*, 15/07/2001: https://www.recyclingtoday.com/article/nonferrous-scrap-terminology/

14 Richard Pallot, 'Amazon destroying millions of items of unsold stock in one of its UK warehouses every year, ITV News investigation finds', ITV News, 21/06/2021: https://www.itv.com/news/2021-06-21/amazon-destroying-millions-of-items-of-unsold-stock-in-one-of-its-uk-warehouses-every-year-itv-news-investigation-finds

15 Thomas Claburn, 'Apple seeks damages from recycling firm that didn't damage its devices: 100,000 iThings "resold" rather than broken up as expected', *The Register*, 05/10/2020.

16 'Burberry burns bags, clothes and perfume worth millions', BBC News, 19/07/2018: https://www.bbc.co.uk/news/business-44885983

17 Christine Frederick, *Selling Mrs. Consumer* (New York: Business Bourse), 1929, p. 246. I first read this quoted in Susan Strasser, *Waste and Want* (New York: Henry Holt and Company), 1999, p. 197.

18 Bernard London, *Ending the Depression Through Planned Obsolescence* (1932): https://upload.wikimedia.org/wikipedia/commons/2/27/London_%281932%29_Ending_the_depression_through_planned_obsolescence.pdf

19 Giles Slade, *Made to Break: Technology and Obsolescence in America* (Cambridge, MA: Harvard University Press), 2006, p. 32.

20 Victor Lebow, *Journal of Retailing* (Spring 1955), p. 7; cited in Vance Packard, *The Waste Makers* (New York: iG Publishing), 1960, p. 38.

21 J. B. MacKinnon, 'The L.E.D. Quandary: Why There's No Such Thing As "Built To Last"', *New Yorker*, 14/07/2016.

22 Slade, *Made to Break*, p. 80.

23 Adam Minter, *Secondhand: Travels in the New Global Garage Sale* (New York: Bloomsbury), 2019, p. 200.

24 Shara Tibken, 'Here's why Apple says it's slowing down older iPhones', *CNET*, 21/12/2017: https://www.cnet.com/tech/mobile/apple-slows-down-older-iphone-battery-issues/

25 Chaim Gartenberg, 'Apple says it could miss $9billion in iPhone sales

due to weak demand', *The Verge*, 02/01/2019: https://www.theverge.com/2019/1/2/18165804/apple-iphone-sales-weak-demand-tim-cook-letter-revised-q1-estimate

26 Chance Miller, 'Apple hit with another lawsuit over iPhone batterygate, seeking $73million in damages for users', *9to5Mac*, 25/01/2021: https://9to5mac.com/2021/01/25/iphone-batterygate-lawsuit-italy/

27 Although the shop owner, Henrik Huseby, never advertised the parts as real, the Norwegian High Court eventually ruled that the screens broke copyright law, and forced Huseby to pay Apple $23,000 in compensation.

28 Brian X. Chen, 'I Tried Apple's Self-Repair Program With My iPhone. Disaster Ensued' *New York Times*, 25/05/2022: https://www.nytimes.com/2022/05/25/technology/personaltech/apple-repair-program-iphone.html

29 Kyle Wiens, 'European Parliament Votes for Right To Repair', iFixit, 25/11/2020: https://www.ifixit.com/News/47111/european-parliament-votes-for-right-to-repair

30 Thomas Claburn, 'Farm machinery giant John Deere plows into two right-to-repair lawsuits', *The Register*, 25/01/2022: https://www.theregister.com/2022/01/25/john_deere_right_to_repair_lawsuits/

31 Fred Lambert, 'Tesla fights back against owners hacking their cars to unlock performance boost', 22/08/2020: https://electrek.co/2020/08/22/tesla-fights-back-against-owners-hacking-unlock-performance-boost/

32 Xia Huo, Lin Peng et al., 'Elevated Blood Lead Levels of Children in Guiyu, an Electronic Waste Recycling Town in China', *Environmental Health Perspectives* 115(7), July 2007, pp. 1113–17: DOI: 10.1289/ehp.9697. I first read about this study in Adam Minter's *Junkyard Planet: Travels in the Billion-Dollar Trash Trade* (New York: Bloomsbury Press), 2013.

33 Bobby Elliott, 'China's Guiyu shifts away from crude processing', *E-Scrap News*, 29/03/2017: https://resource-recycling.com/e scrap/2015/12/17/chinas-guiyu-shifts-away-crude-processing/

34 'Growth in Ghana's port services', *Energy Year*, 06/01/2022: https://theenergyyear.com/articles/growth-in-ghanas-port-services/

35 Josh Lepawsky, 'The changing geography of global trade in electronic discards: time to rethink the e-waste problem', *Geographical Journal* 181(2), June 2015, pp. 147–59: DOI: 10.1111/geoj.12077

36 William Norden, 'Ghana Increases Minimum Wage for 2021 and 2022', *Bloomberg Tax*, 01/09/2021: https://news.bloombergtax.com/payroll/ghana-increases-minimum-wage-for-2021-and-2022

37 Basel Convention E-waste Africa Programme, 'Where are WEEE in Africa?', United Nations Environment Programme, 2011: http://www.basel.int/Implementation/TechnicalAssistance/EWaste/EwasteAfricaProject/Publications/tabid/2553

38 I use 'Agbogbloshie' here to mean the former scrapyard, although in Accra
    the whole neighbourhood – which includes markets and the Old Fadama
    slum – is sometimes called Agbogbloshie, another reflection of how the
    stories we tell can accidentally diminish a place and its people.

39 Kwame Asare Boadu, 'Agbogbloshie redevelopment scheme ready', *Graphic
    Online*, 20/04/2022: https://www.graphic.com.gh/news/general-news/
    ghana-news-agbogbloshie-redevelopment-scheme-ready.html

40 'High Toxic Levels Found at School, Market Neighbouring Informal
    E-waste Salvage Site in Africa', United Nations University, 2011.

41 Jindrich Petrlik, Sam Adu-Kumi et al., 'Persistent Organic Pollutants
    (POPs) in Eggs: Report from Africa', IPEN, 2019: DOI: 10.13140/
    RG.2.2.34124.46723

42 Minter, *Secondhand*, p. 261.

11. THE DAM BREAKS

1 Shasta Darlington, James Glanz et al., 'A Tidal Wave of Mud', *New York
   Times*, 09/02/2019: https://www.nytimes.com/interactive/2019/02/09/
   world/americas/brazil-dam-collapse.html

2 Marcelo Silva De Sousa and Peter Prengaman, 'Arrests in Brazil dam
   disaster, dead fish wash up downstream', Associated Press, 29/01/2019:
   https://apnews.com/article/business-ap-top-news-caribbean-arrests-
   brazil-f8d48cc1ed1e4a7e9a34ac72d384e36e

3 Nathan Lopes, Água do rio Paraopeba apresenta riscos à saúde, diz governo
   de MG', UOL, 31/01/2019: https://noticias.uol.com.br/cotidiano/ultimas-
   noticias/2019/01/31/agua-rio-paraopeba-brumadinho-barragem.htm

4 Marta Nogueira, Tatiana Bautzer, 'Brazil's Vale agrees to \$7 billion
   Brumadinho disaster settlement', Reuters, 04/02/2021: https://www.
   reuters.com/article/us-vale-sa-disaster-agreement-idUSKBN2A41V5

5 Marcos Cristiano Palú, 'Review of Tailings Dam Failures in Brazil',
   conference paper, XXIII Simpósio Brasiliero De Recursos Hídricos, Brazil,
   2019: https://www.researchgate.net/publication/337498009_REVIEW_
   OF_TAILINGS_DAM_FAILURES_IN_BRAZIL

6 Warren Cornwall, 'A Dam Big Problem', *Science*, 21/8/2020: https://www.
   science.org/content/article/catastrophic-failures-raise-alarm-about-dams-
   containing-muddy-mine-wastes

7 Moira Warburton, Sam Hart et al., 'The Looming Risk of Tailings
   Dams', Reuters, 19/12/2019: https://graphics.reuters.com/MINING-
   TAILINGS1/0100B4S72K1/index.html

8 *New York Times*, 'Why Did the Dam in Brazil Collapse? Here's a brief look',
   09/02/2019: https://www.nytimes.com/2019/02/09/world/americas/
   brazil-dam-disaster.html

9   Paulo Trevisani, 'Brazil to Toughen Mine-Safety Rules After Disaster', *Wall Street Journal*, 08/02/2019: https://www.wsj.com/articles/brazils-vale-evacuates-residents-over-safety-fears-at-another-dam-11549622387

10 Dom Phillips, 'Samarco dam collapse: one year on from Brazil's worst environmental disaster', *The Guardian*, 15/10/2016: https://www.theguardian.com/sustainable-business/2016/oct/15/samarco-dam-collapse-brazil-worst-environmental-disaster-bhp-billiton-vale-mining

11 'Report of the International Task Force for Assessing the Baia Mare Accident', United Nations, 2000: https://reliefweb.int/report/hungary/report-international-task-force-assessing-baia-mare-accident

12 Guiomar Calvo, 'Decreasing Ore Grades in Global Metallic Mining: A Theoretical Issue or a Global Reality?', *Resources* 5(4), 2016, p. 36: DOI: 10.3390/resources5040036

13 Alan Septoff, 'How the 20 tons of mine waste per gold ring figure was calculated', *Earthworks*, (2004): https://earthworks.org/publications/how_the_20_tons_of_mine_waste_per_gold_ring_figure_was_calculated/

14 A strange turn of phrase, now that it evokes bougie sourdough bakeries. In this context it reflects the poorest and most rudimentary of mines, beset by pollution and safety issues.

15 'Reducing Mercury Pollution from Artisanal and Small-Scale Gold Mining', United States Environmental Protection Agency: https://www.epa.gov/international-cooperation/reducing-mercury-pollution-artisanal-and-small-scale-gold-mining

16 Maedeh Tayebi-Khorami, Mansour Edraki et al., 'Re-Thinking Mining Waste through an Integrative Approach Led by Circular Economy Aspirations', *Minerals* 9(5), 2019, p. 286: DOI: 10.3390/min9050286

17 Cal Flyn, *Islands of Abandonment: Life in the Post-Human Landscape* (London: William Collins), 2021, pp. 15–41.

18 Dianna Everett, 'Tri-State Lead and Zinc District', *The Encyclopedia of Oklahoma History and Culture*: https://www.okhistory.org/publications/enc/entry?entry=TR014

19 Although *every* bullet is extremely unlikely, the figure was certainly high – it's thought that lead mining in the area amounted to just under 50 per cent of US production at the time.

20 Gerry Blaire, (headline unknown), *Gems and Minerals*, 1968. I read a copy of this article in the Joplin History & Mineral Museum, which holds a thorough archive of mining in the Tri-State area during the twentieth century.

21 John S. Neuberger, Stephen Hu and Rebecca Jim, 'Potential health impacts of heavy-metal exposure at the Tar Creek Superfund site, Ottawa County, Oklahoma', *Environmental Geochemistry and Health* 31(1), 2008, pp. 47–59: DOI: 10.1007/s10653-008-9154-0

22   Perri Zeitz Ruckart, Robert L Jones et al., 'Update of the Blood Lead Reference Value – United States, 2021', Centers for Disease Control and Prevention, 2021: DOI: 10.15585/mmwr.mm7043a4

23   United States Environmental Protection Agency, 'Basic Information about Lead in Drinking Water', 2022: https://www.epa.gov/ground-water-and-drinking-water/basic-information-about-lead-drinking-water

24   Edwin McKnight and Richard Fischer, 'Geology and ore deposits of the Picher Field, Oklahoma and Kansas', United States Geological Survey, 1970.

25   Howard Hu, James Shine and Robert O. Wright, 'The challenge posed to children's health by mixtures of toxic waste: the Tar Creek Superfund Site as a case-study', *Pediatric Clinics of North America* (2007): DOI: 10.1016/j.pcl.2006.11.009

26   Samantha MacBride, *Recycling Reconsidered* (Cambridge, MA: MIT Press), 2013, p. 88.

27   Hristina Byrnes and Thomas C. Frohlich, 'Canada produces the most waste in the world. The US ranks third', *USA Today*, 12/07/2019: https://eu.usatoday.com/story/money/2019/07/12/canada-united-states-worlds-biggest-producers-of-waste/39534923/

28   Credit here to Max Liboiron, who alerted me to this data, although I've used more up to date figures. Max Liboiron, 'Municipal versus Industrial Waste: Questioning the 3-97 ratio', *Discard Studies* (2016): https://discardstudies.com/2016/03/02/municipal-versus-industrial-waste-a-3-97-ratio-or-something-else-entirely/

29   'UK Statistics on Waste', UK Department for Environment, Food & Rural Affairs, 11/05/2022 (accessed: 22/06/22): https://www.gov.uk/government/statistics/uk-waste-data/uk-statistics-on-waste

30   David Cox, 'The planet's prodigious poo problem', *The Guardian*, 25/03/2019: https://www.theguardian.com/news/2019/mar/25/animal-waste-excrement-four-billion-tonnes-dung-poo-faecebook

31   This cleanup appeared to work: later blood tests in 2003 showed that children's blood-lead levels in and around Tar Creek had fallen to near the US average.

32   Ed Keheley, *Chronology of The Tar Creek Superfund Site*, 2017.

33   Leslie Macmillan, 'Uranium Mines Dot Navajo Land, Neglected and Still Perilous', *New York Times*, 31/03/2012: https://www.nytimes.com/2012/04/01/us/uranium-mines-dot-navajo-land-neglected-and-still-perilous.html

34   Michael C. Moncur, 'Acid mine drainage: past, present . . . Future?', *Wat On Earth* by the University Of Waterloo (2006): https://uwaterloo.ca/wat-on-earth/news/acid-mine-drainage-past-presentfuture

35   Stephen Tuffnell, 'Acid drainage: the global environmental crisis you've never heard of', *The Conversation*, 05/08/2017: https://theconversation.com/acid-drainage-the-global-environmental-crisis-youve-never-heard-of-83515

36 'Extent Of The Problem', AbandonedMines.gov (accessed: 23/06/22):
    https://www.abandonedmines.gov/extent_of_the_problem
37 Matthew Brown, '50M gallons of polluted water pours daily from US
    mine sites', Associated Press, 21/02/2019: https://apnews.com/article/
    sd-state-wire-nv-state-wire-north-america-mo-state-wire-in-state-wire-
    8158167fd9ab4cd8966e47a6dd6cbe96
38 Matthew Brown, 'U.S. mining sites dump millions of gallons of toxic waste
    into drinking water sources', Associated Press, 20/02/2019: https://www.
    chicagotribune.com/nation-world/ct-us-mining-wastewater-pollution-
    20190220 story.html

12. HAZARD

1   James Temperton, 'Inside Sellafield: how the UK's most dangerous nuclear
    site is cleaning up its act', *WIRED*, 17/09/2016: https://www.wired.co.uk/
    article/inside-sellafield-nuclear-waste-decommissioning
2   Uranium-based reactors, such as those in the UK's fleet of light-water
    reactors, are the most common, though other fuels such as plutonium and
    thorium are available.
3   'The Cosmic Origins of Uranium', World Nuclear Association, April
    2021: https://world-nuclear.org/information-library/nuclear-fuel-cycle/
    uranium-resources/the-cosmic-origins-of-uranium.asp
4   Nuclear Energy Institute, 2022: https://www.nei.org/about-nei
5   World Nuclear Association, 'Uranium Mining Overview', 2022: https://
    world-nuclear.org/information-library/nuclear-fuel-cycle/mining-of-
    uranium/uranium-mining-overview.aspx
6   Susanne Rust, 'How the US betrayed the Marshall Islands, kindling the
    next nuclear disaster', *Los Angeles Times*, 10/11/2019: https://www.latimes.
    com/projects/marshall-islands-nuclear-testing-sea-level-rise/
7   Cumbria was still Cumberland at that point. 'The Kyshtym Disaster:
    The Biggest Nuclear Disaster You've Never Heard Of', *Mental
    Floss*, 12/11/2015: https://www.mentalfloss.com/article/71026/
    kyshtym-disaster-largest-nuclear-disaster-youve-never-heard
8   'Birds classed as nuclear waste', *The Independent*, 12/03/1998: https://www.
    independent.co.uk/news/birds-classed-as-nuclear-waste-1149723.html
9   Matthew Gunther, 'Stuck in the sludge', *Chemistry World*, 15/09/2015:
    https://www.chemistryworld.com/features/stuck-in-the-sludge/8953.
    article
10 Temperton, 'Inside Sellafield'.
11 UK Health Security Agency, 'Ionising Radiation: Dose Comparisons
    (2011): https://www.gov.uk/government/publications/ionising-radiation-
    dose-comparisons/ionising-radiation-dose-comparisons

12 Paul Brown, 'Huge radioactive leak closes Thorp nuclear plant', *The Guardian*, 09/05/2005: https://www.theguardian.com/society/2005/may/09/environment.nuclearindustry

13 UK Department for Business Energy And Industrial Strategy, '2019 UK Radioactive Material Inventory', 2020: https://ukinventory.nda.gov.uk/wp-content/uploads/2020/01/2019-Materials-Report-Final.pdf

14 Andrew Ward and Alex Barker, 'The nuclear fallout from Brexit, *Financial Times*, 02/03/2017: https://www.ft.com/content/9b99159e-ff2a-11e6-96f8-3700c5664d30

15 The mixture is calcium-zinc instead of borosilicate, if you're interested.

16 UK Defence Select Committee, 'Memorandum submitted by Mr Gordon Thompson', 03/01/2002: https://publications.parliament.uk/pa/cm200102/cmselect/cmdfence/518/518ap02.htm

17 Serhii Plokhy, *Atoms and Ashes: From Bikini Atoll to Fukushima* (London: Allen Lane), 2022, p. 73 (Kindle iOS edition).

18 US Department of Energy, 'What Happened at WIPP in February 2014': https://wipp.energy.gov/wipprecovery-accident-desc.asp

19 Patrick Malone, 'Cost of reviving WIPP after leak could top $500 million', *Santa Fe New Mexican*, 30/09/2014: https://www.santafenewmexican.com/news/local_news/cost-of-reviving-wipp-after-leak-could-top-500-million/article_1402a1fd-9e58-52c6-97df-0241375e839d.html

20 Martin Wainwright, 'Cumbria rejects underground nuclear storage dump', *The Guardian*, 30/01/2013: https://www.theguardian.com/environment/2013/jan/30/cumbria-rejects-underground-nuclear-storage

21 Thomas A. Sebeok, 'Communication Measures To Bridge Ten Millennia', Office of Nuclear Waste Isolation, 1984: https://www.osti.gov/servlets/purl/6705990

22 'Und in alle Ewigkeit: Kommunikation über 10 000 Jahre: Wie sagen wir unsern Kindeskindern wo der Atommüll liegt?', *Journal of Semiotics* (1994).

23 Angelique Chrisafis, 'France to build up to 14 new nuclear reactors by 2050, says Macron', *The Guardian*, 10/02/2022: https://www.theguardian.com/world/2022/feb/10/france-to-build-up-to-14-new-nuclear-reactors-by-2050-says-macron

24 Hans Von Der Burchard, 'Germany to extend runtime of two nuclear plants as "emergency reserve"', *Politico*, 05/09/2022: https://www.politico.eu/article/germany-to-extend-two-runtime-of-two-nuclear-plants-as-emergency-reserve/

25 Dan Murtaugh and Krystal Chia, 'China's Climate Goals Hinge on a $440 Billion Nuclear Buildout', *Bloomberg*, 02/11/2021: https://www.bloomberg.com/news/features/2021-11-02/china-climate-goals-hinge-on-440-billion-nuclear-power-plan-to-rival-u-s

26 'Energy strategy: UK plans eight new nuclear reactors to boost production', BBC News, 07/04/2022: https://www.bbc.co.uk/news/business-61010605

EPILOGUE: PRECIOUS

1   Heather Rogers, *Gone Tomorrow: The Hidden Life of Garbage* (New York: The New Press), 2005, p. 3.
2   Ministry of Environment and Food of Denmark, *Life Cycle Assessment of grocery carrier bags* (Copenhagen: The Danish Environmental Protection Agency), 2018: https://www2.mst.dk/Udgiv/publications/2018/02/978-87-93614-73-4.pdf
3   Robin Murray, *Zero Waste* (London: Greenpeace), 2002, p. 18.
4   Michelle Slatalla, 'A Visit From The Priestess of Waste-Free Living', *New York Times*, 15/02/2010: https://www.nytimes.com/2010/02/16/fashion/18spy.html
5   Stephen Moss, 'The zero-waste revolution: how a new wave of shops could end excess packaging', *The Guardian*, 21/04/2019: https://www.theguardian.com/environment/2019/apr/21/the-zero-waste-revolution-how-a-new-wave-of-shops-could-end-excess-packaging
6   Madeleine Cuff, 'People aren't coming anymore: refill stores face crisis as shoppers lose interest in zero-waste lifestyles', *I News*, 17/03/2022: https://inews.co.uk/news/zero-waste-shopping-crisis-refill-stores-1522978
7   Samantha MacBride, *Recycling Reconsidered* (Cambridge, MA: MIT Press), 2013, p. 6.
8   Prevented Ocean Plastic Research Centre, 2022: https://www.preventedoceanplastic.com/what-is-prevented-ocean-plastic
9   For example, Baolong Ma, Xiaofei Li et al., 'Recycling more, waste more? When recycling efforts increase resource consumption', *Journal of Cleaner Production* (2018): DOI: 10.1016/j.jclepro.2018.09.063
10 I first read about this study in Adam Minter's *Junkyard Planet: Travels in the Billion-Dollar Trash Trade* (New York: Bloomsbury Press), 2013. See: Jesse R. Catlin and Yitong Wang, 'Recycling gone bad: When the option to recycle increases resource consumption', *Journal of Consumer Psychology* 23(1), 2013: DOI: 10.1016/j.jcps.2012.04.001
11 Cited in Adam Minter, *Secondhand: Travels in the New Global Garage Sale* (New York: Bloomsbury), 2019, p. xiii.
12 Neil Lancefield, 'Millions of garages too full of junk to store cars – survey', *London Evening Standard*, 28/06/2021: https://www.standard.co.uk/news/uk/rac-b942895.html
13 I feel somewhat better after reading life-cycle analyses, which suggest that reusables only become more sustainable by the second child.
14 Scottish Government, 'Scotland's deposit return scheme', 2021: https://www.gov.scot/news/scotlands-deposit-return-scheme/
15 'Historic day in the campaign to beat plastic pollution: Nations commit to develop a legally binding agreement', United Nations Environmental Program, 02/03/2022: https://www.unep.org/news-and-stories/

press-release/historic-day-campaign-beat-plastic-pollution-nations-commit-develop

16 Tim Burrows, 'The only grave is Essex: how the county became London's dumping ground', *The Guardian*, 25/10/2016: https://www.theguardian.com/cities/2016/oct/25/london-dumping-ground-essex-skeleton-crossrail-closet

17 Rachel Salvidge and Jamie Carpenter, 'MAPPED: England and Wales' toxic legacy landfills', *Ends Report*, 15/01/2021: https://www.endsreport.com/article/1704522/mapped-england-wales-toxic-legacy-landfills

18 I'm here counting 'industrial waste' as unknown content, because the term describes the creator, not its constituents. Rachel Salvidge, 'Toxic waste lies beneath schools in England and Wales, map shows', *The Guardian*, 15/01/2021: https://www.theguardian.com/environment/2021/jan/15/toxic-waste-lies-beneath-schools-and-homes-uk-landfill-map-shows

19 Robert J. Nicholls, Richard P. Beaven et al., 'Coastal Landfills and Rising Sea Levels: A Challenge for the 21st Century', *Frontiers in Marine Science* (2021): DOI: 10.3389/fmars.2021.710342

20 Rebecca Morelle, '"Shocking" levels of PCB chemicals in UK killer whale Lulu', BBC News, 02/05/2017: https://www.bbc.co.uk/news/science-environment-39738582

INDEX